Geology of the Canyon Ferry Quadrangle of Montana

U.S. Dept. of Interior

with an introduction by Kerby Jackson

Introduction

It has been almost a century since the Department of Interior released it's important publication "Geology of the Canyon Ferry Quadrangle, Montana". First released in 1951, this important volume has been out of print and has been unavailable to the mining community since those days, with the exception of expensive original collector's copies and poorly produced digital editions.

It has often been said that *"gold is where you find it"*, but even beginning prospectors understand that their chances for finding something of value in the earth or in the streams of the Golden West are dramatically increased by going back to those places where gold and other minerals were once mined by our forerunners. Despite this, much of the contemporary information on local mining history that is currently available is mostly a result of mere local folklore and persistent rumors of major strikes, the details and facts of which, have long been distorted. Long gone are the old timers and with them, the days of first hand knowledge of the mines of the area and how they operated. Also long gone are most of their notes, their assay reports, their mine maps and personal scrapbooks, along with most of the surveys and reports that were performed for them by private and government geologists. Even published books such as this one are often retired to the local landfill or backyard burn pile by the descendents of those old timers and disappear at an alarming rate. Despite the fact that we live in the so-called "Information Age" where information is supposedly only the push of a button on a keyboard away, true insight into mining properties remains illusive and hard to come by, even to those of us who seek out this sort of information as if our lives depend upon it. Without this type of information readily available to the average independent miner, there is little hope that our metal mining industry will ever recover.

This important volume and others like it, are being presented in their entirety again, in the hope that the average prospector will no longer stumble through the overgrown hills and the tailing strewn creeks without being well informed enough to have a chance to succeed at his ventures.

Kerby Jackson
Josephine County, Oregon
November 2016

CONTENTS

CONTENTS

CONTENTS

GEOLOGY OF THE CANYON FERRY QUADRANGLE

By John B. Mertie, Jr., Richard P. Fischer, and S. Warren Hobbs

ABSTRACTS

The Canyon Ferry quadrangle lies between meridians 111°30′ and 111°45′ and parallels 46°30′ and 46°45′ in west-central Montana. As a new dam on the Missouri River is to be built at Canyon Ferry, the Reclamation Service wished to obtain all available information regarding the geology and mineral resources of this quadrangle. With funds furnished by the Reclamation Service, the accompanying geologic map and report were made in 1946 by the Geological Survey.

The Missouri River flows northwest across the quadrangle, and an old dam at Canyon Ferry impounds a lake 6 miles long, called Lake Sewell. Two tributaries of the river from the northeast, and one from the southwest, are perennial; all the others are intermittent throughout most of their courses. The Big Belt Mountains constitute the northeastern watershed; and the Spokane Hills, though not a principal watershed, bound the valley on its southwest side. Prominent river-cut terraces at several levels are present.

This region has a semiarid climate, characterized by short, warm summers and long, cold winters. The mean annual precipitation—about 12 inches—falls mainly in May, June, and July; and a cyclic variation exists, with minima every 15 years. Vegetation and animal life are scant. Canyon Ferry is the only settlement in the quadrangle.

The Reclamation Service plans to irrigate large tracts of land in the valleys of Jefferson, Madison and Gallatin Rivers, the headwaters of the Missouri. Much of the water used for this purpose will never return to the river; and to compensate for these losses, larger storage facilities must be provided. A new dam therefore will be built at Canyon Ferry, with a power plant for generating 67,000 kilowatts.

The sedimentary rocks that are mapped consist of 5 pre-Cambrian formations, 11 Paleozoic formations, 5 Tertiary units, and gravels of Pleistocene and Recent age. The pre-Cambrian formations comprise the Newland limestone, Greyson shale, Spokane shale, Empire shale, and Helena limestone, all belonging to the Belt series. An unknown thickness of overlying Belt rocks has been removed by erosion, so that the oldest Cambrian rocks rest unconformably on the Helena limestone.

The Paleozoic rocks comprise six formations of Cambrian age, two of Devonian age, two of Carboniferous age, and one of Permian age. The Cambrian formations, named from oldest to youngest, are the Flathead quartzite, Wolsey shale, Meagher limestone, Park shale, Pilgrim limestone, and Dry Creek shale. The Park and Dry Creek shales have not everywhere been separately mapped, but instead have been included respectively with the Meagher and Pilgrim limestones.

1

The upper part of the Meagher limestone is dolomitic. The Devonian formations are the Jefferson limestone, which is dolomitic, and the overlying Three Forks shale. The Carboniferous units, named from oldest to youngest, are lower and upper units of the Madison limestone, and the lower and upper units of the Quadrant formation. The Permian formation is tentatively designated as the Phosphoria formation.

The Tertiary beds consist of four mappable units and one unit of undifferentiated sediments. All of them are lake beds, composed in whole or in part of volcanic debris. The oldest unit is conglomeratic; the next younger contains beds of bentonite, and one or more beds of diatomaceous earth; the third is largely reworked volcanic detritus; and the youngest consists of conglomerate, grit, and some limestone. Less consolidated beds of Miocene age are present just outside the quadrangle.

, Pleistocene gravels occur as bench deposits at three principal levels, approximately at 250, 150, and 50 feet above the present level of the Missouri River; but other ancient gravels also are present at intermediate altitudes. Glaciation of pre-Wisconsin age is believed to have preceded the deposition of the oldest gravels; no Wisconsin glaciation has been recognized in the area. Recent gravel, sand, and silt constitute the alluvium of the valley floor.

The igneous rocks comprise monzonitic and granitic intrusives with felsic and mafic differentiates; pyroxene diorite and gabbro; latitic intrusives that may be related to the monzonitic rocks; andesitic intrusives, commonly greenstones; and lava flows and dikes of basalt. The monzonitic and granitic rocks are believed to be of early Tertiary age and the lavas are probably of Miocene age; the ages of the other igneous rocks are indeterminate.

The principal industries are farming and stock raising, but gold placers and gold lodes were mined in earlier years, and a small amount of gold placer mining is still being done. Descriptions are given of the principal sites, past and present, of gold mining. Many deposits of copper ores have been located, but only one was ever brought to the stage of production. A small amount of platinum metals was produced as a byproduct of gold placer mining. The nonmetalliferous resources include dimension stone, road metal, concrete aggregate, riprap, ballast, lime, cement rock, dolomite, bentonite, and industrial sapphires; only sapphires have been produced commercially.

The production of placer gold from this quadrangle and its vicinity is estimated to have been $17,500,000; of lode gold, $600,000; of copper, $501,000; of sapphires, $75,000; and of platinum metals between 150 and 200 ounces.

INTRODUCTION

LOCATION OF AREA

The greater Helena mining region is defined by Pardee and Schrader (1933, p. 1) as an area of about 3,000 square miles, surrounding the city of Helena, in west-central Montana. The Canyon Ferry quadrangle, bounded by meridians 111°30′ and 111°45′, and by parallels 46°30′ and 46°45′ constitutes an area of about 205 square miles in the eastern part of the greater Helena mining region. About 61 square miles of the northwest part of the quadrangle lies in Lewis and Clark County; the remainder is a part of Broadwater County. This quadrangle, though not yet formally named, takes its tentative designation from the small settlement of Canyon Ferry, which lies near its western edge, about 16 miles N. 78° E. of the business section of Helena.

EARLIER SURVEYS

A bibliography of reports describing the geology and mineral resources of the greater Helena mining region has been presented by Pardee and Schrader (1933, p. 8). The principal geologic publications that apply directly to the Canyon Ferry quadrangle are the following:—

> Walcott, Charles D., Pre-Cambrian fossiliferous formations: Geol. Soc. America Bull., vol. 10 pp. 199–244, 1899.
>
> Douglass, Earl., New vertebrates from the Montana Tertiary: Carnegie Mus. Annals, vol. 2, pp. 145–200, 1903.
>
> Walcott, Charles D., Algonkian formations of northwestern Montana: Geol. Soc. America Bull., vol. 17, pp. 1–18, 1906.
>
> Douglass, Earl, Some new merycoidodonts: Carnegie Mus. Annals, vol. 4, pp. 99–109, 1907.
>
> Douglass, Earl., Fossil horses from North Dakota and Montana: Carnegie Mus. Annals, vol. 4, pp. 267–277, 1908.
>
> Jennings, O. E., Fossil plants from the beds of volcanic ash near Missoula, western Montana: Carnegie Mus. Annals, vol. 8, no. 2, pp. 385–427, 1920.
>
> Pardee, J. T., Geology and ground-water resources of Townsend Valley, Montana: U. S. Geol. Survey Water-Supply Paper 539, 1925.
>
> Pardee, J. T. and Schrader, F. C., Metalliferous deposits of the greater Helena mining region: U. S. Geol. Survey Bull. 842, 1933.

Earlier engineering surveys constitute the geodetic control for the base map on which the accompanying geologic map, plate 1, is charted. The triangulation was done by the Missouri River Commission, the Forest Service and the General Land Office. Plans and profile surveys of the Missouri River, prepared by the Geological Survey (Marshall, 1914, pl. 1–F) in 1913, show its position and character at that time. Airplane photographs, from which the base map was compiled, and on which the geologic boundaries were drawn in the field, were supplied by the Agricultural Adjustment Agency and the Forest Service. The land net shown on plate 1 is based mainly on cadastral surveys made by the General Land Office. In 1945 and 1946 the Reclamation Service prepared a topographic map in 6 sheets showing the terrain along the Missouri River from the site of the newly proposed Canyon Ferry dam upstream to Townsend. This map is drawn to a scale of 1 : 12,000, with contour intervals of 5 and 10 feet, with a maximum altitude of 3,800 feet. The land net along the river, relocated at the same time, is shown also. The altitudes mentioned in this report are based on first-order and second-order leveling done by the Coast and Geodetic Survey, and on third-order altitudes supplied by the Forest Service.

PRESENT WORK

The building of a new Canyon Ferry Dam, proposed in 1944[1] to replace the older dam built about 50 years ago, supplied a cogent need for this report. The Reclamation Service wished to obtain all possible information regarding the geology and mineral resources of the Canyon Ferry quadrangle, wherein the new reservoir will be situated; and to this end requested the Geological Survey to prepare a geologic map and report on this quadrangle.

The personnel assigned to this work in 1946 included Richard P. Fischer, S. Warren Hobbs, and John B. Mertie, Jr., assisted by Stanley E. Good and Peter Joralemon. Fischer spent 4½ months in field work; Hobbs, 3½ months; Mertie, 4¾ months; Good, 3¼ months; and Joralemon, 3 months. A mineral resources "appraisal party", consisting of Robert M. Dreyer assisted by Alfred L. Bush, spent about 1½ months in and about Helena in a study of the nonmetalliferous resources. Dreyer and Bush mapped unit 2, of the Tertiary sequence.

The base map on which the geologic map is plotted was prepared by the Topographic Division of the Geological Survey under the direction of Mr. Jesse Mundine. The base map was prepared on a scale of 1:31,680, which therefore became the scale of geologic compilation. The geodetic control consisted of 41 locations within or lying close to the Canyon Ferry quadrangle: 7 were established by the Missouri River Commission, 2 by the General Land Office, and 32 by third-order triangulation done by the Forest Service. The aerial photographs, from which the base map was prepared, were made with a 8¼-inch lens from an elevation of about 13,750 feet above the ground to produce a scale of approximately 1:20,000. These photographs had overlaps of 50 percent or more so that the geologists were able, by the use of folding pocket stereoscopes, to obtain three-dimensional imagery in plotting geologic contacts.

The geologic map was prepared by Fischer, Hobbs, and Mertie: Fischer mapped the hard-rock geology of the southeastern Big Belt Mountains and the Spokane Hills, Hobbs mapped the northwestern part of the Big Belt Mountains, and Mertie mapped the Tertiary rocks and alluvial deposits. Fischer and Hobbs contributed notes on the pre-Cambrian and Paleozoic rocks; Dreyer and Bush prepared a statement on the bentonite and other nonmetalliferous deposits of the quadrangle, parts of which were used in the present report; and Mertie obtained the available information on gold placers and certain other deposits of economic significance. The report was written by Mertie.

---

[1] Missouri River Basin: Conservation, control, and use of water resources. 78th Cong., 2d sess., S. Doc. 191, presented by Mr. O'Mahoney on April 12, 1944. Washington, 1944.

ACKNOWLEDGMENTS

The writer gratefully acknowledges the special effort put forth by the Agricultural Adjustment Agency in Salt Lake City and by the Forest Service in Missoula, in preparing prints of aerial photographs for use by the beginning of the field season. Thanks are due also to the Missoula office of the Forest Service for furnishing a large number of third-order positions that were needed for geodetic control in the Canyon Ferry quadrangle. The Reclamation Service furnished the topographic map described above, and cooperated fully in all possible ways. The writers acknowledge with thanks the information regarding the older placer mining operations, given by Roland C. Eames, of Helena, William H. De Borde, of Oregon Gulch, C. J. Sheriff, of Canyon Ferry, and Owen Perry, of the Perry-Schroeder Mining Co., Helena. Thanks are due to I. G. Sohn of the U. S. Geological Survey for drafting the geologic map.

GEOGRAPHY

DRAINAGE

The Missouri River, master stream of the region, flows about N. 40° W. across the Canyon Ferry quadrangle, dividing it into a southwestern triangle having an area of about 55 square miles, and a northeastern polygon with an area of approximately 150 square miles. At Canyon Ferry a thirty-foot dam, built in 1896–98, impounds a reservoir known as Lake Sewell, at an altitude of about 3,678 feet. The altitude of water level at the Winston bridge across the Missouri is about 3,715 feet, whereas it is 3,650 feet at the foot of the Canyon Ferry Dam. Therefore, in an airline distance of 12 miles, the original fall of the river was about 5 feet to the mile. Lake Sewell, whose length is between 5 and 6 miles, conceals the original course of the Missouri in this stretch, but it is evident that the river crowded the west wall of its valley throughout the quadrangle.

The drainage pattern is unsymmetrical, as nearly all the larger tributaries enter from the northeast. The principal southwest-trending tributary valleys are those of Confederate Creek, White Gulch, Bilk Gulch, Avalanche Creek, Hellgate Gulch, Little Hellgate Gulch, Magpie Creek, Cave Gulch, Oregon Gulch, and in the northwest corner of the quadrangle, Trout Creek. Horse Gulch and Clark Gulch are major tributaries of Oregon Gulch; and York Gulch, with its tributary Kingsbury Gulch, are important branches of Trout Creek. Confederate and Trout Creeks are perennial streams, but all the others are intermittent, though White Gulch, Avalanche Creek, Hellgate Gulch, and Magpie Creek are commonly perennial within the Big Belt Mountains. The only east-flowing tributary of any size is Beaver

Creek, a perennial stream which heads in the Elkhorn Mountains and crosses the southern end of the quadrangle, emptying into the Missouri River about 2 miles upstream from the mouth of White Gulch. Spokane Creek drains the western slopes of the Spokane Hills and enters the Missouri River beyond the northwestern limit of the Canyon Ferry quadrangle.

Certain features of the tributaries entering from the northeast merit special mention. The headwater courses of all of them trend from south to southwest, so that they appear as a back-hand drainage with respect to the Missouri. Farther downstream, on the foreland southwest of the Big Belt Mountains, the back-hand drainage is less evident. This feature, in the absence of structural control, suggests that the stream that preceded the present Missouri River may have drained southeastward during some earlier physiographic cycle.

The gradients of Confederate, Beaver, and Trout Creeks are adjusted to the present base level of the Missouri, but the streams of smaller size are only partly adjusted. Thus streams of intermediate size, such as White Gulch, Avalanche Creek, Hellgate Gulch, and Magpie Creek, and Cave, Oregon, and Clarks Gulches, have headwater gradients that are adjusted to an ancient base level, oversteepened gradients farther downstream, and gradients in their lower courses that are adjusted to the latest base level. Still smaller streams show lack of adjustment even in their lower courses. An extreme example is a broad valley floor, known as Dry Hollow, which lies northwest of the lower valley of Confederate Creek, and clearly was a former outlet of that stream to the Missouri.

Beaver Creek, issuing from the Elkhorn Mountains, originally flowed southeastward in the direction of Townsend, or more probably discharged at the east side of these mountains into some master stream that flowed southeastward, draining the Prickly Pear Valley. The present lower valley of Beaver Creek is a new outlet to the Missouri, produced by stream piracy or by superposition.

The Missouri River and its tributaries in western Montana show many physiographic anomalies that have not been satisfactorily explained. One of them is the broad open valley of the Missouri River upstream from Canyon Ferry, with a succession of gorges and open stretches downstream that made possible the Canyon Ferry, Hauser, and Holter dams. From the mouth of Magpie Creek downstream for about a mile, the Missouri originally followed an old course northeast of Canyon Ferry, impinging against the gravel bluffs at the mouth of Cave Gulch. A later superposition of the river onto its western wall has formed a short gorge, at the lower end of which the present Canyon Ferry dam is located.

RELIEF

MOUNTAINS AND VALLEYS

The Big Belt Mountains on the east and the Elkhorn Mountains on the west bound this part of the Missouri River Valley. The crest of the Big Belt Mountains, at the heads of Avalanche and Magpie Creeks, is a rolling upland that rises between 3,000 and 4,000 feet above the valley floor of the Missouri River. The tributaries of the river, however, are deeply incised, and the southwestern limit of the mountains is very abrupt so that the northeastern wall of the valley appears precipitous. From the southwest face of the Big Belt Mountains a gravel-covered foreland, ranging in width from 1 to 6 miles, slopes gently toward the river or to Lake Sewell. This foreland, which appears to be a gravel-covered pediment, has a complex geomorphic history as it was originally the site of an ancient lake but later became the main valley of the Missouri. Owing to its low altitude and the terraces carved upon it, this foreland is regarded as a part of the valley floor. The Spokane Hills, between the Big Belt Mountains and the Elkhorn Mountains, constitute an intermediate bounding divide that rises to an altitude of 1,500 feet above the Missouri River. The northeastern lateral spurs of the Spokane Hills descend abruptly to the valley floor.

The valley of the Missouri River, between the fronts of the Big Belt Mountains and the Spokane Hills, is a basin with a maximum width of about 6 miles. It forms one arm of a much longer and wider depression, called Townsend Valley, that extends from Canyon Ferry upstream for 35 miles to Toston. The Prickly Pear Valley, another arm of this depression, lies west of the Spokane Hills and was probably at one time the main valley of the Missouri. The Spokane Hills constitute in reality a hard-rock peninsula that separates these two arms of Townsend Valley. The Townsend and Prickly Pear Valleys, considered as a unit, exemplify a number of extensive basins, separated from one another by constricted courses of the river, that characterizes the headwaters of the Missouri. The present bed of the river is slightly incised in the older valley floor of Townsend Valley.

TERRACES

Numerous terraces, which mark older base levels at which the Missouri River and its tributaries successively flowed, are well developed. Terraces have longitudinal slopes that depend principally on the older stream gradients, and lateral slopes that depend on several causes. In order to correlate such erosion levels throughout an area, the older stream gradients must be determined and projected from place to place, and for such an undertaking a detailed topographic map is indispensable.   No such map has yet been made of the Canyon Ferry quadrangle, and the correlations mentioned below therefore must be regarded as tentative and subject to revision.

Pardee (1925, p. 6–8) has divided these bench-like forms into three major groups, on the basis of their relative ages.   The oldest, and highest, of these terraces has altitudes ranging from 250 to 300 feet above the present level of the Missouri River; the second, or next younger terrace, has altitudes from 150 to 200 feet above the river; and the third, or lowest, terrace is about 50 feet above the river level. Besides these terrace levels, numerous less well developed erosion levels exist, but their correlation at present is impracticable.

The best example of the first or oldest terrace is the sloping gravel-covered surface that extends from the south end of the Spokane Hills southeastward beyond the Canyon Ferry quadrangle toward the Missouri River.   Beaver Creek transects this surface, but from the hills on either side of the creek, its original continuity is apparent.   The diversion of Beaver Creek to its present course occurred at the time when this stream, or the master stream into which it discharged, flowed at the level of this old erosion surface.   Below this level, but above that of the second terrace, an intermediate level is imperfectly developed at a few places along the south side of Beaver Creek.

The second of the more distinct terraces is well developed in the valley of Beaver Creek, and at many other sites within the quadrangle. On Beaver Creek it shows as a wide flat on both sides of the creek, sloping gently toward the Missouri.   Along the northeast side of the Missouri River, the second terrace forms the upper surface of the foreland that extends southwestward from the face of the Big Belt Mountains.   Between Confederate Creek and Avalanche Creek, it is everywhere present and shows at numerous places as a low escarpment rising above one or more lower terraces, but nowhere is it close enough to the Missouri to form any high escarpment.   Farther downstream, however, the second terrace is well developed at two bluffs close to Lake Sewell, one just northwest of the lower course of Hellgate Gulch, and another, known as Magpie Bluff, which is just southeast of the lower course of Magpie Creek.   Along the top of the spur west of Cave Gulch, the terrace is developed close to the face of the Big Belt Mountains.   Farther up the river, beyond the limits of the quadrangle, the second terrace also is visible along the southwest wall of the Missouri.

The lowest of the well-marked terraces, known as terrace 3, lies at an altitude of about 50 feet above the mean water level of the Missouri. It is shown along both sides of the valley of Beaver Creek, though it is less well developed than the second terrace.   It is believed that this erosion surface may be correlated with the low one along the east side of the Elkhorn Mountains, contiguous to the valley of the ancient stream that discharged toward Townsend. It may be correlated also with the well-developed low terrace that marks the eastern wall of Spokane Creek, along the west flank of the Spokane Hills.

The lowest terrace is well marked along Confederate Creek, and also along the northeast bank of the Missouri River downstream from Dry Hollow. At these two localities, however, an erosion level below terrace 2, and about 50 feet above terrace 3, also shows to advantage. The valley floor of Dry Hollow, which is cut at the level of terrace 3, connects with the river through a short steep gorge. The unadjusted valley of Dry Hollow shows that Confederate Creek discharged through this outlet at a time when the base level of the Missouri was about 50 feet higher than at present. The lowest terrace is also intermittently developed southwest of the Missouri River, upstream from the limits of the quadrangle.

The lowest terrace shows in most of the larger valleys, from White Gulch to Canyon Ferry. Another erosion level, very imperfectly developed in the lower valley of Avalanche Creek, lies below terrace 3. The latter, and other less perfectly developed erosion levels, are recognizable on the lateral spurs descending into Hellgate Gulch and Magpie Creek, and in Cave Gulch. Near the mouth of Cave Gulch, northwest of the junction of the Cave Gulch road and York road, is a thick bank of gravel resting on the eroded surface of tilted Tertiary beds. This surface is about 50 feet above the present level of the Missouri River, and unquestionably correlates with the lowest or third terrace. It dates the stage of erosion when the Missouri flowed northeast of Canyon Ferry and impinged against these gravel bluffs at the lower end of Cave Gulch, and it serves to correlate this stage with the development of Dry Hollow.

CLIMATE

Climatic records have been kept at Canyon Ferry and at Helena by the U. S. Weather Bureau for many years. At Canyon Ferry the record began in 1899, and at Helena in 1880; these records have been summarized by the U. S. Weather Bureau (1936) up to 1930. The tabulation of mean temperatures, given below, is taken from this 1930 compilation, but the mean precipitation and snowfall, which are of greater importance for a reclamation or power project, has been supplemented and brought up to date by including the annual records of the Weather Bureau from 1931 to 1945.

Climatic Data at Canyon Ferry and Helena, Montana

[Records of U. S. Weather Bureau]

	Length of record (years)	January	February	March	April	May	June	July	August	September	October	November	December	Annual
Average monthly temperature (°F.):														
Canyon Ferry	32	19.4	23.2	33.3	44.6	53.0	61.2	68.4	65.9	56.0	44.2	32.3	21.0	43.6
Helena	51	20.2	23.6	32.6	43.7	51.6	59.6	67.3	66.1	55.6	45.3	32.9	24.8	43.7
Average maximum temperature (°F.):														
Canyon Ferry	32	28.6	34.0	45.0	58.0	66.6	75.7	84.7	82.0	69.5	56.6	42.0	30.0	56.1
Helena	51	28.4	32.2	41.8	54.0	62.2	71.1	80.3	79.1	67.1	55.3	41.3	32.4	53.8
Average minimum temperature (°F.):														
Canyon Ferry	32	10.1	12.3	21.6	31.3	39.4	46.8	52.2	49.8	42.5	31.8	22.6	12.1	31.0
Helena	51	12.1	15.0	23.4	33.3	41.0	48.2	54.2	53.1	44.1	35.4	24.4	16.9	33.4
Average monthly precipitation (in.):														
Canyon Ferry	47	.41	.41	.47	.81	1.90	2.20	1.24	.93	1.13	.73	.44	.47	11.14
Helena	66	.79	.59	.75	1.01	1.95	2.27	1.10	.79	1.19	.85	.66	.73	12.68
Average monthly snowfall (in.):														
Canyon Ferry	45	6.4	5.1	4.5	0.9	0.1	T	0.0	0.0	T	1.0	3.7	5.9	27.6
Helena	66	10.2	8.4	9.3	5.4	1.9	T	T	T	0.9	3.8	6.5	8.6	55.0

The climate of this part of the Missouri River Valley, at altitudes between 3,500 and 4,000 feet, is characterized by short warm summers and long cold winters, with the precipitation of a semiarid country. The midday temperatures in June, July, and August are high, with a maximum as high as 104°, and a mean maximum ranging from 76.8° at Helena, to 80.8° at Canyon Ferry; but the mean minimum temperatures for the three summer months (51.8° at Helena and 49.6° at Canyon Ferry) show that the nights are cool. The mean minimum temperatures for the months of December, January, and February are 14.7° at Helena and 11.5° at Canyon Ferry. The extreme range in temperature is from 104° in summer to −42° in winter, a total of 146°. The humidity, however, is fairly low, with day to night means from April to September of 39 to 66 percent, and similar means from October to March of 60 to 71 percent. The average dates of the first and last killing frosts at Helena are September 29 and May 7; and at Canyon Ferry, September 22 and May 13.

The distribution of precipitation and snowfall throughout the year at Canyon Ferry and Helena is given in the preceding tabulation, and is shown graphically in figure 1. In this chart, snowfall is converted to water by the conventional assumption that 10 inches of snow is equal to 1 inch of water. The true ratio in this area is closer to 13

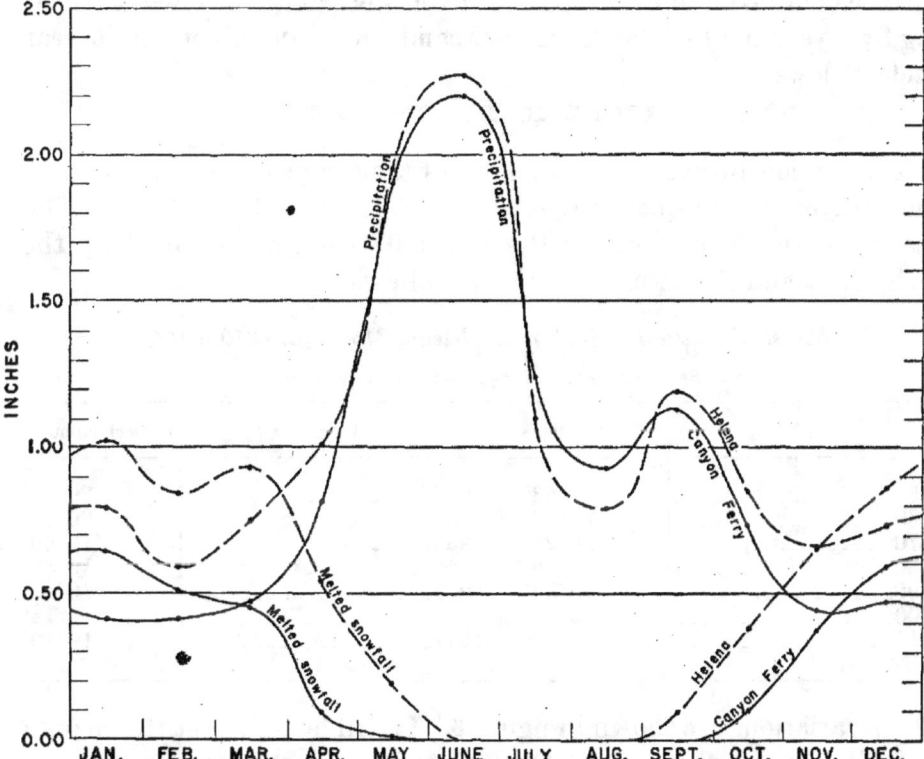

FIGURE 1.—Monthly precipitation and snowfall, Canyon Ferry and Helena.

921852—51——2

to 1. The maximum rainfall is in June, and the maximum snowfall in December, but these are average values and dates from which great departures occur. The total mean precipitation at Helena is about 14 percent greater than at Canyon Ferry, and the mean snowfall is about twice as great.

A periodic variation in the annual precipitation was stated by Pardee (1925, p. 12) to exist with a half-period of 9 years, but 25 years of additional observations are now available beyond those to which Pardee had access. To test this periodicity, the mean annual precipitation for Canyon Ferry and Helena is charted in figure 2, from the time of the earliest records to 1945. The maxima, or peaks of the wet periods, are not well defined, but from 1891 to 1908 to 1925 a period of 17 years is suggested. The minima, however, which show the peaks of the dry periods, are better defined, and from 1889 to 1904 to 1919 to 1935 indicate a period of 15 years. Most of the rainfall in this area comes from summer thunderstorms, which are erratically distributed. It is therefore too much to expect that the records of precipitation at one or two stations will show perfect cyclic variation, even if such cycles were really well defined. A better interpretation of climatic periodicity could be obtained by averaging the records from many stations; and as an index of the compensation that might thus be obtained, the 1934 low at Canyon Ferry may be cited, correcting by 1 year the 1935 low at Helena, and producing a perfect 15-year cycle of lows.

MAGNETIC DECLINATION

The average horizontal declination of the compass, for the center of the Canyon Ferry quadrangle, during 1946 was 19°14' east. The record of this variation for the last 100 years, as supplied by the U. S. Coast and Geodetic Survey,[2] is as follows:

East Magnetic Declination at Helena, Montana, 1850–1947

[From records of U. S. Coast and Geodetic Survey]

Year	Declination	Year	Declination
1850	19°53'	1920	20°34'
1860	20°10'	1930	20°03'
1870	20°21'	1935	19°50'
1880	20°10'	1940	19°36'
1890	19°54'	1945	19°17'
1900	20°00'	1946	19°14'
1910	20°33'	1947	19°10'

These variations are shown in figure 3. It will be seen that the change in declination has been about 1⅓ degrees in the last 27 years, or about 1 degree in 20 years.

[2] Letter responding to a request by telephone.

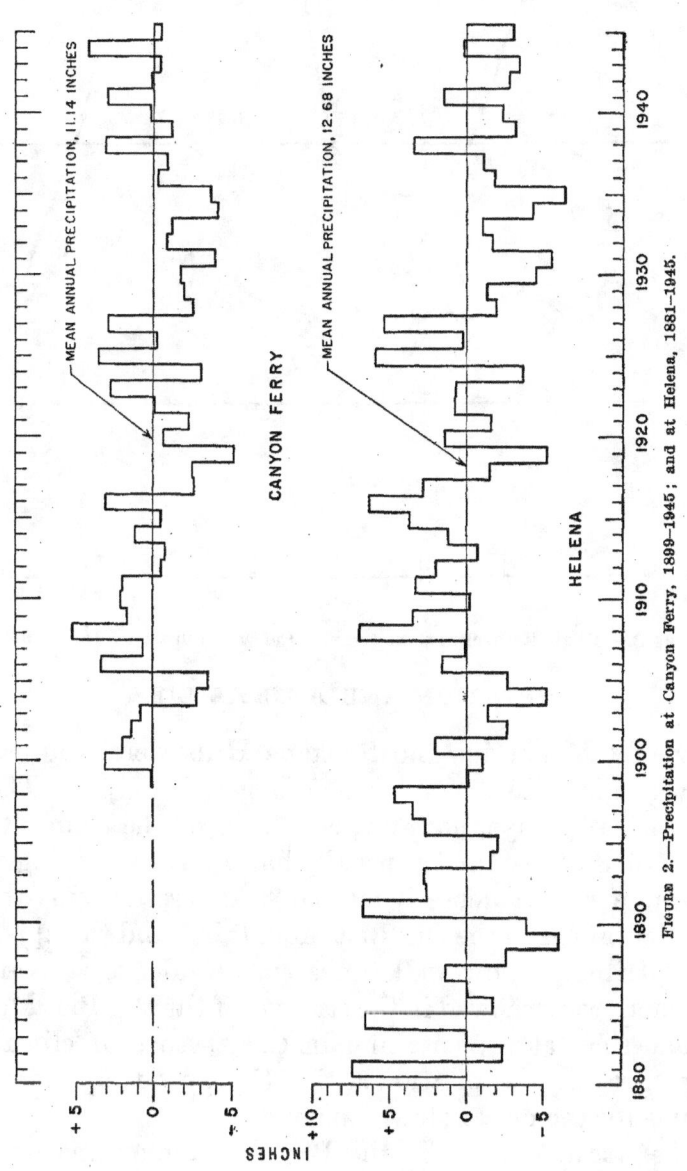

FIGURE 2.—Precipitation at Canyon Ferry, 1899–1945; and at Helena, 1881–1945.

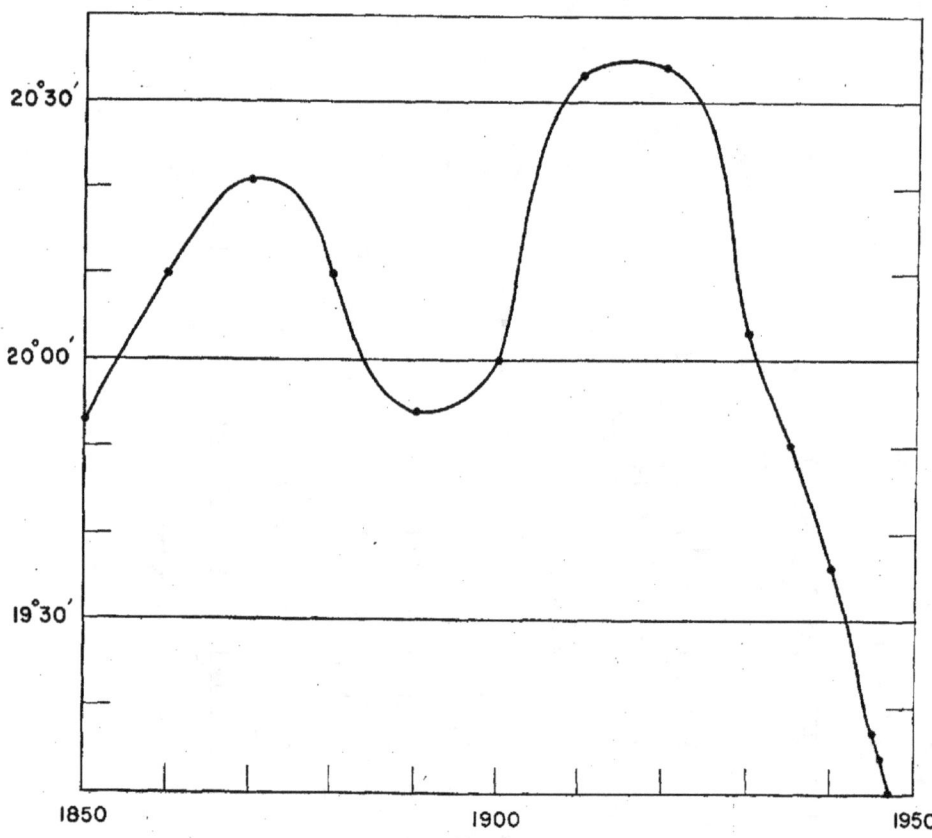

FIGURE 3.—Magnetic declination, Canyon Ferry quadrangle, 1850–1947.

VEGETATION AND ANIMAL LIFE

The Big Belt Mountains and Spokane Hills were originally well timbered with pine and fir, and some cedar and juniper. The placer miners of the early days, however, cut much of the timber that was accessible to their camps; and as usually happens near mining camps, a great deal more was accidentally burned. As a result, only the upper reaches of the valleys in the Big Belt Mountains, and the upper slopes of Spokane Hills are now well timbered. Evidence is available to prove that more water flowed in the streams of the Big Belt Mountains 80 years ago than at present; and in the absence of climatic data indicating a greater precipitation at that time, probably deforestation is responsible for the present lower amount.

Along the stream courses in the Big Belt Mountains is a sparse growth of willow brush, poplar and aspen, grasses, and other vegetation that grow near moisture in such an environment. Beyond the mountains, however, the vegetation is sagebrush, tumbleweeds, cactus, bunch-grass, and other plants that are characteristic of the semiarid West.

Few animals live in this region. Deer are common in the higher ranges, and antelope are fairly common, both in the hills and in the lower country. A few black bear live in the Big Belt Mountains, and some beaver are found along the perennial streams. Coyotes and ground squirrels, and rabbits are fairly common. Rattlesnakes are not restricted to any one section, but are most common in the low arid hills north of the lower part of Beaver Creek. Black snakes are probably as numerous as rattlesnakes in the Big Belt Mountains, but are less prevalent in the lower country.

SETTLEMENTS

Canyon Ferry, founded by John Oakes in 1865, is the only settlement in the Canyon Ferry quadrangle. The population consists of several farmers and their families, a school teacher and family, the manager of the power plant and his family, and several government employees and their families. East Helena is the nearest source of supplies.

Several towns, now entirely abandoned, once existed in or close to this quadrangle when gold placer mining was a thriving enterprise. Among them were Cavetown in the valley of Cave Gulch; Jimtown at the head of Clarks Gulch; White City on White Gulch; Diamond City on Confederate Creek; and French Bar on the Missouri River. Soon after the discovery of gold on Confederate Creek, Diamond City became a town with a population of 5,000 persons; and White City, soon after its founding, had a population of 1,000.

TRANSPORTATION

The highway connecting Three Forks with Helena passes across the southwest corner of the Canyon Ferry quadrangle, and makes readily accessible the western flanks of the Spokane Hills. Three secondary roads lead eastward from this highway to three bridges crossing the Missouri River, and connect with other secondary roads along the northeast side of the river. The northernmost of these roads crosses the Missouri at the mouth of Trout Creek and continues on to York, and a branch from this road leads southward via Clarks Gulch and Cooper Gulch to Canyon Ferry. The second side road leads from East Helena direct to Canyon Ferry. The third secondary road leads eastward from Winston down the valley of Beaver Creek to a bridge across the Missouri, whence it connects with the network of secondary roads on the northeast side of the river. Five tertiary, but passable, roads cross the Big Belt Mountains, within or close to this quadrangle. They follow up the valleys of White Gulch, and Trout, Magpie, Avalanche, and Confederate Creeks, and connect with similar roads on the east side of the Big Belt Mountains. Local roads of the same type also go some distances up the valleys of Cave, Hellgate, and Little Hellgate Gulches.

INDUSTRIES

This part of Montana was settled first by prospectors, who in 1864 discovered gold on Last Chance Creek at the present site of Helena, and on Confederate Creek. A narrative of these and other discoveries is given in a later chapter; but it suffices here to state that a gold placer mining industry was soon established which, in its waning stages, included some mining of gold and copper lodes. Placer mining was still in progress near Helena in 1946, on Indian Creek west of Townsend, and on White Gulch; and new work was being planned for Confederate Creek. Mining, however, can no longer be considered an important industry in or close to the Canyon Ferry quadrangle, though changing economic conditions may rejuvenate this industry to some degree in coming years.

Farming and stock raising are now the principal industries of the Canyon Ferry quadrangle. Farming is done mainly on the floor of Townsend Valley, but 8 miles of this valley floor is already submerged by Lake Sewell, and 14 additional miles will be flooded eventually by the new Canyon Ferry dam. Most of this farming is done by means of irrigation, as the area is too dry to raise crops consecutively by dry farming only. Some dry farming is done, but is successful only during years or cycles of excess precipitation.

The generation of hydroelectric power may be regarded as an industry in this quadrangle and will become still more important after the new Canyon Ferry dam is constructed. Three important dams have been built on the Missouri River in the 28-mile stretch downstream from Canyon Ferry. The smallest and oldest of them is the Canyon Ferry dam, which was completed in 1898, and now generates 6,700 kilowatts of power. The Hauser dam, about 12 miles downstream, was completed in 1907, was washed out in 1908, and was rebuilt in 1911. At this site 17,000 kilowatts of power are generated. The Holter dam, about 28 miles downstream from Canyon Ferry, was built about 1918 and generates 38,400 kilowatts of power. These three and certain other hydroelectric plants farther downstream are privately owned and operated.

The Reclamation Service plans to irrigate large tracts of land in the valleys of the Jefferson, Madison, and Gallatin Rivers, the headwater tributaries of the Missouri which join at Three Forks. New dams will be built for this purpose and the water from one older dam, now used for power, will be diverted to irrigation. Much of this water will never return to the river, for which reason large losses of water will result. The Reclamation Service therefore was obligated to provide larger storage facilities to maintain the present flow of water. The new Canyon Ferry dam will be constructed primarily for this purpose, though a small part of its water will be pumped to irrigate

26,600 acres in the Helena Valley. It will be built 6,900 feet downstream from the present dam, and its power plant will generate 67,000 kilowatts.

The new Canyon Ferry dam will impound water to an altitude of 3,800 feet, and will flood the Townsend Valley to the mouth of Indian Creek, about 1½ miles downstream from Townsend. Large tracts of fertile bottom land that are now cultivated thus will be destroyed, and some additional land adjoining the reservoir will become unusable. Moreover, no new land will be irrigated along the terraces northeast of the river, as the water is to be used mainly for storage and power. A new Holter dam, downstream from the present one, would have given an equivalent storage of water, and would have submerged no agricultural land; but such a project would have flooded the present Holter and Hauser power installations. The over-all cost of a new Holter dam was therefore regarded as excessive, and the project was not developed.

GENERAL GEOLOGY

SEDIMENTARY ROCKS

The Canyon Ferry quadrangle shows an incomplete section of pre-Cambrian and Paleozoic rocks, no recognizable rocks of Mesozoic age, a great thickness of Tertiary beds, and Quaternary gravels of several ages. Five pre-Cambrian formations, belonging to the Belt series, have been recognized and mapped. The Paleozoic sequence comprises six formations of Cambrian age, two of Devonian age, and five units of Carboniferous and Permian age, all of which are separately mapped where this is feasible. At some localities, however, poor exposures made it necessary to map two formations as a single unit.

The Oligocene beds are fresh-water deposits of lacustrine and fluviatile origin that have been divided into four mappable units. Another group, probably also of Oligocene age, whose stratigraphic relations to the mapped units of this age cannot be exactly determined, is shown as undifferentiated Tertiary. Miocene beds occur at two localities, one just south and the other just north of this quadrangle.

The Quaternary deposits have a long and involved history that is incompletely understood. Pleistocene gravels of several ages could be distinguished and mapped, but without the aid of a topographic map they are grouped as a single unit. The Recent gravels of the valley floors are separately shown.

PRE-CAMBRIAN ROCKS

BELT SERIES

The Belt series consists of thick units of shale, siltstone, sandstone, and limestone, with an aggregate thickness within this quadrangle of more than 8,000 feet.   The series is divided into five formations based on lithology and structure, but each formation grades into adjoining ones, so that their boundaries are to some extent a matter of interpretation.   Most of the rocks in the Belt series are thinly and evenly bedded.

The rocks of the Belt series are well exposed but do not commonly form bold outcrops, except along steep valley walls.   On the rolling upland that forms the crest of the Big Belt Mountains, these rocks crop out inconspicuously; but at many places along the west slope of the Spokane Hills, the Spokane and Empire shales have been eroded to form smooth gentle slopes with very few outcrops.

An erosional unconformity at the base of the Cambrian rocks transects the two uppermost pre-Cambrian formations, so that their total thickness is not exposed.   Recent work by Clyde S. Ross,[3] moreover, has demonstrated that the pre-Cambrian rocks of the Big Belt Mountains are only the lower part of the Belt series of Montana.

NEWLAND LIMESTONE

The Newland limestone is the oldest pre-Cambrian formation that occurs in this quadrangle, but it is not the basal member of the Belt series.   It crops out in the northeastern part of the quadrangle, in the central part of the Big Belt Mountains.   Its base is not exposed, but its local thickness is estimated to be about 2,000 feet.

The Newland is a uniformly dense, dark-gray, dolomitic limestone, thinly and evenly bedded, except in its upper part.   It crops out at few places, but forms abundant float of brown platy fragments that resemble a hard fissile shale.   Thin beds of blue-gray limestone are common in the upper part of the formation, and the uppermost 200 feet consists of bands of such limestone, separated by zones from 20 to 40 feet thick of alternating thin beds of limestone and shale. These uppermost beds constitute a transitional sedimentary facies that marks the top of the Newland limestone and the base of the overlying Greyson shale.   The contact is arbitrarily drawn on the highest recognizable bed of limestone.   At some places these uppermost beds show considerable deformation, consisting of minor folds, contortions, and even faulting, but such deformation cannot be traced far into adjacent beds of shale.

GREYSON SHALE

The Greyson shale crops out in a broad, northwest-trending band in the Big Belt Mountains.   It is well exposed along Avalanche, Magpie,

---

[3] Ross, Clyde S., oral communication.

and Trout Creeks, but it forms few good exposures on the divides between the valleys. Its thickness is between 2,000 and 3,000 feet.

The formation consists mainly of shale, siltstone, and fine-grained sandstone, all of which are thinly and evenly bedded. The color of these rocks is dark gray to dark brown, but they weather to sombre shades of brown or red. Medium- to coarse-grained sandstones, in places somewhat conglomeratic, are present in the upper few hundred feet of the formation. They are dominantly light brown or light gray, but partly white or red. Some of these beds show cross-bedding and ripple marks, indicative of deposition in shallow water.

The Greyson shale grades upward through a transitional zone about 200 feet thick into the overlying Spokane shale. These transitional rocks consist of red shale of the Spokane type interbedded with light-colored sandstone. The contact between the two formations is drawn at the top of the uppermost recognizable bed of sandstone.

SPOKANE SHALE

The Spokane shale crops out with good exposures in the Big Belt Mountains from the vicinity of White Gulch along the east side of the quadrangle northward to Trout Creek. It crops out also on the west slope of the Spokane Hills, which is the type locality, as established by Walcott in 1899. The thickness is between 1,500 and 2,000 feet.

The Spokane consists mainly of soft, poorly bedded shale, and some siltstone is present in the lower part of the formation. Bright red is the dominant color, but a part of the rock is altered to a greenish color in small spots, bands, or along zones several tens of feet thick. In some places this greenish coloration appears to be related to fractures or igneous intrusions; but in other places it is uncertain whether the rock was originally red or green and what caused a change, if such occurred, from one color to the other. For this reason, the criterion of red versus green color, which in the Big Belt Mountains is the principal difference between the Spokane and Empire shales, may not hold over any extensive area.

The Spokane shale, though soft, is tough and resistant to weathering, so that it forms rather prominent outcrops. It breaks down into small chips and flakes that resist abrasion, and are transported for considerable distances in the stream beds. Detritus of such material is particularly noticeable in the uppermost unit of the Tertiary beds.

EMPIRE SHALE

The Empire shale is exposed from the eastern side of the quadrangle, in the vicinity of White Gulch, to the vicinity of Oregon Gulch, where it pinches out; it is present also along the west and north sides of the Spokane Hills. The formation has a maximum thickness of about 1,000 feet in the vicinity of Hellgate canyon and is about 800 feet thick on the west side of the Spokane Hills. Northwest of Magpie Creek and on the west side of the Spokane Hills, where the Helena limestone is absent, part of the Empire was probably eroded before deposition of the Flathead quartzite, of Cambrian age.

The Empire is commonly a hard, dense, siliceous shale or argillite that is thinly bedded and banded. The color ranges from light to dark shades of greenish gray but red shale like that in the Spokane is abundant in its lower part. Some thin beds of limestone are interlayered with the siliceous shale in the upper part of the Empire. Most of them are fine-grained to dense, blue-gray beds, but those near the top are light-brownish gray and closely resemble the limestone of the overlying Helena limestone.

The contact between the Empire and Spokane shales is gradational and is the most difficult contact in the Belt series to establish and map. At most places it is arbitrarily drawn at the base of the lowest greenish, siliceous shale that characterizes the Empire, but in a few places some of this shale is included in the Spokane. The Empire shale, as mapped, thus may include in its lower part a considerable amount of red shale of the Spokane type; and the contact shown on the map is therefore consistent only within an interval of about 200 feet. The contact of the Empire and Helena also is gradational, with rock types characteristic of each formation alternating at the contact, but only through a stratigraphic interval of 25 to 50 feet.

HELENA LIMESTONE

The Helena limestone is exposed only along the western part of the Big Belt Mountains. It is about 500 feet thick at Hellgate Gulch and possibly a little thicker in the vicinity of White Gulch; but owing to an erosional unconformity which transects the uppermost pre-Cambrian rocks at a low angle, this formation thins to a feather edge in the valley of Magpie Creek.

The Helena is composed of thinly but unevenly bedded, commonly laminated, fine-grained to dense dolomitic limestone. Most of the beds are a small fraction of an inch thick, but some are as much as 2 inches thick, and a few are a foot thick. Thin partings of shaly limestone or limy shale occur between some of the beds of dolomite.

The Helena limestone does not form conspicuous bluffs, but numerous small exposures are common. The dolomite disintegrates as hard platy layers, so that the area where this dolomite crops out is strewn with such platy fragments that weather tan to reddish brown.

PALEOZOIC ROCKS

CAMBRIAN SYSTEM

FLATHEAD QUARTZITE

The Flathead quartzite crops out along the southwestern slopes of the Big Belt Mountains and along the crest of the Spokane Hills; it is exposed also in the headwater part of Avalanche Creek in the northeastern corner of the quadrangle. In the Big Belt Mountains, this quartzite forms a single hogback ridge southeast of Hellgate Gulch; but northwest of that stream it is involved in the folding that predominates in that part of the quadrangle, forming repeated hogbacks and synclinal canoe-shaped ridges.

The quartzite acts mainly as a brittle unit and is displaced by numerous small cross faults which do not appreciably affect the adjoining formations. Locally these cross faults are so numerous as to produce jagged, steplike outcrops of the formation. Some are of mappable dimensions, as in the area south of Hellgate Gulch, and in the central part of the Spokane Hills; but at numerous other places the displacements are too small to be mapped on the scale of the accompanying map. Northwest of Hellgate Gulch, the Flathead is tightly folded and bent, but shows little rupturing. The average thickness is about 200 feet.

The Flathead is composed of medium-grained to rather coursegrained quartzite, part of which is finely conglomeratic. The beds of conglomerate are irregularly distributed throughout the formation, both stratigraphically and along the strike. No basal conglomerate is present. The quartzite is dominantly light colored, mostly pale gray; but purple-red banding is common and conspicuous. It weathers tan to brown, but much of the weathered surface is covered with lichens, which give the rock a dark appearance when viewed from a distance. Bedding is well developed, mostly in layers 1 to several feet thick; but some of the quartzite is thin-bedded and some is cross-bedded.

The contact of the Flathead quartzite with the Belt series is everywhere covered with talus; it is obviously an unconformable one as it truncates the Helena limestone and part of the Empire shale, though at a very low angle. Similarly the contact with the overlying Wolsey shale is covered with talus in most places, but poor exposures in the Spokane Hills suggest that the Flathead grades upward into the Wolsey.

WOLSEY SHALE

The Wolsey shale is exposed in the central part of the Big Belt Mountains at the head of Avalanche Creek, along the southwestern slope of the mountains from White Gulch to Trout Creek, and along the crest of the Spokane Hills. As the formation is soft and lies between resistant strata, strike valleys and saddles form on it. Outcrops are poor, but its presence generally can be established by float. The Wolsey varies greatly in thickness as a result of deformation, but its average thickness is about 400 feet.

The Wolsey is composed mainly of fine-grained sandy shale, and thin discontinuous layers of sandstone. Most bedding surfaces of the sandy beds are crinkly, and show mud cracks, worm borings and casts, and in addition minor wrinkles probably resulting from deformation. The bedding surfaces are partly covered with films of greenish shale and abundant mica, which imparts to the weathered rock a bright sheen, resembling that of a phyllite. The sandstone is brown, but the associated shale is gray-green and gives a greenish color to most of the rock. A little red sandstone and shale is present in places.

Limy shale and thin and irregular beds of limestone also form a part of the Wolsey shale, particularly in the upper horizon. The limestone is partly fine-grained and blue-gray, and partly medium-grained and gray, tinged with green and red. Glauconite grains are common in some of the crystalline limestone, and oolite occurs rarely.

In the northern part of the Spokane Hills, near the Canyon Ferry stock, the Wolsey strata are considerably altered, both by thermal metamorphism, and by the injection of Tertiary igneous material. By thermal metamorphism the shale is baked to a dark-colored hornfels, the sandy beds are changed to dark quartzite, and the limy layers are changed to a dense, hard, siliceous, light- to dark-green rock. At several sites in the Spokane Hills, however, the shale shows lit-par-lit and other forms of igneous injection; and at some places the rock is so altered that it is mapped as igneous.

The Wolsey shale is thought to grade upward into the overlying Meagher limestone, with an increase in limestone and a decrease in shale. Exposures of this gradational zone, however, are uncommon and generally poor.

MEAGHER LIMESTONE

The Meagher limestone is well exposed along the southwestern slope of the Big Belt Mountains from White Gulch to Trout Creek, and in the central part of the mountains at the head of Avalanche Creek. It forms bold outcrops in cliff faces and along ridges. It is well exposed also along the crest of the Spokane Hills, though its outcrops are subdued compared with those in the Big Belt Mountains.

The Meagher in the Big Belt Mountains is a fine-grained, gray limestone that weathers light gray to blue gray. From a distance it appears to be a massive rock, but actually it is thinly though unevenly bedded in layers from 1 to 3 inches thick. On weathering, however, it does not separate into slabs along these bedding planes. Some beds are medium-grained and granular, representing possibly recrystallized oolites, or a sand of fossil fragments. Near the top and base of the formation are thin discontinuous layers of light gray to bright yellow and bright red limestone. Commonly this bright-colored limestone occurs also as small "chips" in the gray limestone, thus suggesting an intraformational conglomerate. The Meagher in the Spokane Hills, close to the Canyon Ferry stock, is recrystallized to a coarse-grained, white marble. The thin-bedded character is less conspicuous in the Spokane Hills.

The Meagher limestone has reacted to disastrophism in somewhat the same way as the Flathead quartzite, though to a lesser degree. It appears as a resistant member of the sedimentary series, and is commonly fractured by cross faults. Owing to its massive character and its wide distribution, it forms prominent outcrops, which outline better than any other limestone the structure of the Paleozoic rocks. In the area of tight asymmetrical folding near the mouth of Magpie Creek, it is involved in an overturned anticline, whose axial plane is a reverse fault. Adjacent to this anticline is a syncline that is overturned in the opposite direction. In such areas of complex folding, the thickness is inconstant, but in general is about 300 feet.

PARK SHALE

The Park Shale, which lies above the Meagher limestone and below the Pilgrim limestone, occurs both in the Spokane Hills and in the Big Belt Mountains, but is everywhere poorly exposed. Its outcrop at most places can be mapped because shallow strike valleys and saddles are formed on it. Along the southwestern slope of the Big Belt Mountains, however, it has not everywhere been differentiated from the Meagher limestone, as it rarely crops out and its topographic expression is not striking. Shale belonging to this formation was not recognized at the head of Avalanche Creek.

The Park shale in the Spokane Hills is commonly gray to greenish, and weathers greenish or olive drab, or is stained brown with limonite. It is fairly well-bedded, and thin-bedded, somewhat micaceous, and in part silty, with subordinate fine-grained sandstone layers. Near the Canyon Ferry stock it has been altered to a hard, black shale.

The Park Shale in the Big Belt Mountains is gray, thin-bedded, and finely silty. It weathers in places to splinter-shaped fragments and pencil slate, which cover the surface of the ground, and in the absence of outcrops, serve to identify the formation. Pseudomorphic crystals of limonite after pyrite are common in places.

The Park shale, because it is soft and incompetent, and because it lies between two massive limestones, is much deformed. The production of splinter shale seems to be related to tight squeezing near the crests of folds. Owing to the structure the thickness varies greatly, ranging from a knife edge to nearly twice its usual thickness. The thickness in the Spokane Hills is probably about 300 feet; but along the southwestern slopes of the Big Belt Mountains the thickness appears to be less than 100 feet. Nowhere in the quadrangle are exposures good enough to determine its structural relation to the overlying and underlying limestones, but it is believed to be conformable with both.

PILGRIM LIMESTONE

The Pilgrim limestone forms bold outcrops along the crest of the Spokane Hills, where it is about 500 feet thick. It is well exposed in the faces of cliffs and along ridges in the southwestern slope of the Big Belt Mountains and at the head of Avalanche Creek. Where measured in Hellgate canyon the formation is faulted, so its true thickness could not be determined; but it is thought to have about the same thickness in the mountains as in the Spokane Hills, even though the lithology differs appreciably in the two areas.

The Pilgrim limestone in the Spokane Hills is a fine-grained to medium-grained dolomitic limestone of granular texture. It is dominantly light gray, with a darker gray mottling, and at some places white (bleached?) areas. Some of it, notably the upper part, is white. Most of the limestone is rather massive and poorly bedded, though near the top it is well bedded and in places thin-bedded. The basal part is thinly bedded and contains platy layers and intraformational conglomerates of light-gray, red, and yellow limestone.

The Pilgrim limestone in the Big Belt Mountains can be divided conveniently into two units. The lower, thicker unit is rather uniform in general character though varied in detail. It is composed of some dense or fine-grained limestone, either dark gray or white; some rather coarse-grained limestone; and some "muddy" limestone, which is light gray, yellow, or red in color and forms irregular and discontinuous layers and intraformational conglomerates. These beds weather buff, yellow, or reddish brown. All of them are thin-bedded, and many have a jagged or hackly surface. They grade into the upper unit, which consists of light-gray medium-grained dolomitic limestone of granular texture. This limestone occurs in rather well-bedded strata, one to several feet thick, and contains no intraformational conglomerate, though locally it shows coarse and irregular mottling.

DRY CREEK SHALE

No good exposures of the Dry Creek shale exist in the quadrangle, though its outcrop makes a narrow but fairly continuous bench on the eastern slope of the Spokane Hills, where it is distinguished on the map from the underlying Pilgrim limestone. Outcrops of the Dry Creek shale were recognized only in a few isolated places in the Big Belt Mountains, and the formation is differentiated on the map from the Pilgrim limestone only at a few places.

The shale is soft, brownish, and somewhat sandy and micaceous. Its average thickness in the Spokane Hills is about 50 feet, and it is certainly no thicker than that in the Big Belt Mountains.

DEVONIAN SYSTEM

JEFFERSON LIMESTONE

The Jefferson limestone occurs in the Big Belt Mountains at the head of Avalanche Creek, and along the southwestern slope of the mountains from the vicinity of Avalanche Creek to Trout Creek. It commonly crops out as ridges and hogbacks, making it easy to recognize; and it also produces a characteristically blocky type of dark brownish-gray float that serves to identify it where outcrops are lacking. It is well exposed along the eastern slope of the Spokane Hills, though it makes no bold outcrops. The formation is about 400 feet thick in Hellgate canyon, and about 500 feet thick in the Spokane Hills.

In the Big Belt Mountains the Jefferson limestone is a dark gray to brown, dolomitic limestone, part of which has a strongly fetid or oily odor when freshly broken. The rock is medium-grained, with a granular texture, and weathers to a felted or sugary surface. Bedding is pronounced and even, mostly in layers from a few inches to a few feet thick, and part of the limestone is thinly banded or laminated, though it shows little or no tendency to weather along these planes. A little light gray, brown-weathering chert is present along some bedding planes. Some of the limestone beds are brecciated and consist of small fragments that are firmly cemented with a matrix the same as or similar to the fragments. They are more easily seen on a weathered surface than on a fresh one. As no definite relation between this brecciation and regional deformation is apparent, the breccia may possibly be of sedimentary origin.

All of the above-mentioned lithologic features, except color, apply to the Jefferson limestone in the Spokane Hills, where it is either blue black, weathering black, or pale-gray weathering white. The contact between limestones having these contrasting colors generally follows the beds, but at places it crosses the bedding, the blue-black color fading into the pale gray through an interval about 1 foot thick. This color contrast is less pronounced some distance away from exposed igneous rocks, and possibly is related to thermal alteration.

The Jefferson limestone folds without shattering, and acts as a resistant rock forming the crests of synclinal ridges. Northwest of Hellgate Gulch, four such synclines occur within a distance of 7 miles. The contacts of the Jefferson limestone with the underlying Dry Creek shale and the overlying Three Forks shale were not seen.

THREE FORKS SHALE

The Three Forks shale crops out in the southeastern part of the Spokane Hills, and though exposures are rather poor, it makes easily recognized strike valleys and saddles. It is about 300 feet thick. Along the southwestern slope of the Big Belt Mountains the outcrop was recognized and mapped separately from the adjoining formations in only a few places; elsewhere it is included with the Jefferson limestone. The thickness of the Three Forks in the Big Belt Mountains was not determined by measurement but is probably less than 300 feet.

The Three Forks is composed mainly of drab to purple-gray, poorly bedded shale. It weathers dark gray to brown, with abundant limonitic stains in places. Pseudomorphic crystals of limonite after pyrite are common. Some of the shale is finely sanded and slightly micaceous and some is calcareous, in part forming shaly limestone that contains poorly preserved fossils. A minor amount of black, thinly laminated shale occurs with the siliceous shale. In the Spokane Hills the upper 30 to 50 feet of the formation consists dominantly of thinly bedded, dense, and hard siliceous shale of light-gray, creamy, or pale-green color. As the outcrops are distinct and produce easily recognized float, this formation is a useful horizon marker.

CARBONIFEROUS SYSTEM

MADISON LIMESTONE

The Madison limestone crops out in the southeastern part of the Spokane Hills, along the front of the Big Belt Mountains from Little Hellgate Gulch southeast to Bilk Gulch, in the vicinity of White Gulch, and at the head of Avalanche Creek. It forms many bold, cliff-forming outcrops, such as those along the mountain front in Hellgate Canyon and in the canyon of Avalanche Creek. No completely un-faulted section of the formation exists in the quadrangle. The Madison limestone has been divided into two units, which are separately mapped.

The lower unit of the Madison limestone is thin-bedded, and in most places is readily distinguished from the upper, more massive unit by this character alone. Beds more than a foot thick are rare in the lower unit and most of them are only a few inches thick. The color of the limestone in the Big Belt Mountains is dominantly light, mostly gray, but shades of blue, brown, and red are present in places, and some of the rock is dark gray, and has a faint oily odor when freshly broken. The limestone commonly weathers cream or buff, but in some places is reddish brown. Equivalent beds in the Spokane Hills are blue gray except near large igneous intrusions, where they are altered to white marble. The granularity of the unaltered limestone ranges from dense to coarse. Some of the beds, especially those that are coarse-grained, are fossiliferous, many consisting of fragments of shells and crinoid stems. A little light-colored chert is present in places. At the head of Avalanche Creek a minor amount of soft, fine-grained, thin-bedded, yellow and reddish sandstone occurs with beds of fossiliferous limestone.

Part of the upper unit of the Madison is massive and poorly bedded; part is thinly and evenly bedded but weathers as a massive rock. The limestone is dominantly light gray, but some of it is dark gray and has a fetid or oily odor when freshly broken. Except in the Spokane Hills, where the upper Madison everywhere has been altered to a coarsely crystalline marble, the granularity ranges from dense through fine-grained to coarse-grained limestone; some of the dense limestone has an almost vitreous or waxy luster. The coarse-grained rock commonly shows abundant fragments of fossils, but specimens well enough preserved for identification are rare.

One collection of fossil invertebrates was made from the Madison limestone. The locality, collector, and determination of the fossils by James S. Williams, are given below:

46 F 72 c. Site near the crest of a south-trending spur in the NE1/4 sec. 5, T. 5 N., R. 14 E., about half a mile north of Avalanche Gulch, at an altitude of 6,200 feet. Collector, Richard P. Fischer.

This fauna is a characteristic Madison limestone fauna, containing such characteristic species as *Spirifer centronatus* Winchell, *Chonetes logani* Norwood and Pratton, *Leptaena analoga* (Phillips), *Chonetes loganensis* Hall and Whitfield, and other species generally found associated with these.

The lower and upper limestone units of the Madison limestone are probably to be correlated with two similar and persistent formations of Mississippian age, the Lodgepole and Mission Canyon limestones, which are typically exposed in and near the Little Rocky Mountains, in north-central Montana about 200 miles east-northeast of the Canyon Ferry quadrangle.

The contact of the Madison limestone with the underlying Three Forks shale is not exposed, and the upper contact of the Madison was not definitely recognized, though possibly a normal contact with the Quadrant formation is present in the southeastern part of the Spokane Hills. The total thickness of the Madison is about 1,000 feet.

QUADRANT FORMATION

The Quadrant formation is exposed along the front of the Big Belt Mountains in the vicinity of White Gulch and at places in the southeastern part of the Spokane Hills. Two units have been recognized and mapped between White Gulch and Bilk Gulch, and for a short distance southeast of White Gulch. These units are distinguished also at the southeastern end of the Spokane Hills. Elsewhere the recognition of separate units is not feasible. Beds included in the Quadrant consist of quartzite, in part interbedded with limestone, and sandstone and shale. These beds have probably an aggregate thickness of several hundred feet, but are too poorly exposed and are too much deformed by faulting and folding to establish accurately their true thickness and stratigraphic sequence.

The quartzite of the Quadrant is hard, tough, and brittle, and is dense to glassy in appearance. It is predominantly light gray and light pink, weathering brownish. Bedding is obscure except where the quartzite is interbedded with limestone. The limestone is fine-grained, gray to brownish gray, weathering light gray, and is fairly well bedded, with irregular thin cherty layers along some bedding planes. The sandstone is thin bedded, and is brown, red, or gray. Most of it is soft and shaly, but some is quartzitic and some is limy. Impressions of plant fossils show along some of the bedding planes. The shales are mostly sandy and red, though some purple, green, and gray shale are present.

PERMIAN SYSTEM

PHOSPHORIA (?) FORMATION

A group of rocks of doubtful age is exposed along the front of the Big Belt Mountains in the vicinity of White Gulch, and for a short distance to the south. Several hundred feet of these beds crop out in the valley of White Gulch, but as the formation is bounded by faults, its total thickness cannot be determined. Tentatively these rocks are assigned to the Phosphoria formation.

The formation is composed of limestone, shale, and a subordinate amount of sandstone. All of the limestone is thin-bedded and most of it is shaly. Brownish gray, weathering to light brown, is the predominant color, but some of it is blue gray and some is black with a strong oily odor. Black chert occurs with some of the limestone beds, particularly those that are black. The granularity of the limestone ranges from fine to coarse, the latter commonly containing poorly preserved fossils. The shale is poorly exposed but is probably fairly abundant. It is brown to light gray and weathers brown; some is limy and some is silty. The sandstone is brown to red, fine-grained, and thin-bedded.

One collection of fossil invertebrates was made from the rocks here described as a part of the Phosphoria(?) formation. The locality, collector, and determination of these fossils by James S. Williams and Helen Duncan, are given below:

46 F 25. Small mining prospect in SE¼NE¼ sec. 19, T. 10 N., R. 2 E., just south of the road that leads up White Gulch, and about half a mile east of the mountain front. Collector, Richard P. Fischer.

This collection consists of one individual horn coral, several bryozoans, and many pieces of brachiopods. The coral has been tentatively referred to the genus *Bradyphyllum*, and the bryozoans to the genera *Septopora*, *Fenestella*, *Eridopora*(?), and *Rhombopora*. There is also a stenoporoid bryozoan. Except for a new species of *Productus*, *sensu lato*, a single specimen referable to *Linoproductus phosphaticus* (Girty), another specimen that may be an Aulosteges, and some Compositas, the brachiopods, though common, are represented by incomplete and macerated specimens that are not definitely identifiable. The brachiopods, in general appearance, resemble a Phosphoria assemblage; the bryozoans, though not diagnostic, are agreeable to a Phosphoria age assignment; and *Linoproductus phosphaticus* (Girty) is a Phosphoria species. The evidence is not conclusive, but the factors cited above make a Phosphoria age assignment a plausible one.

CENOZOIC ROCKS

TERTIARY SYSTEM

GENERAL CHARACTER

The headwater part of the valley of the Missouri River, and some of its tributary valleys, are characterized by a number of wide basins that are the sites of former Tertiary lakes. After these lakes disappeared, their basins were aggraded and filled with sediments of various types, including much material of volcanic origin. One of these basins, heretofore described as the Townsend Valley, is the site of a great sequence of Tertiary beds, largely of volcanic origin, which in the Canyon Ferry quadrangle occupy the lowland between the Spokane Hills and the southwestern front of the Big Belt Mountains. Tertiary beds are present also southwest of the Spokane Hills, within the adjacent basin known as the Prickly Pear Valley.

The Tertiary sediments are composed dominantly of tuffaceous materials, but in various degrees are mixed with nonvolcanic materials derived from the nearby hills. These sediments were deposited mainly in ponds, but prior to final deposition, many of these materials, both volcanic and nonvolcanic, were reworked and transported by streams within the basins of Townsend and Prickly Pear Valleys. The total thickness of these sediments is between 6,000 and 10,000 feet.

Pardee (1925, pp. 17–21) has listed and summarized all the paleontological data that bear on the age and mode of deposition of these Tertiary beds, both in this area and in other of the Missouri basins. The general consensus of all this information indicates that the oldest known Tertiary lake beds in the Canyon Ferry quadrangle are of

Oligocene age. The fossils so far found suggest that beds of Miocene age are present at other localities close to this quadrangle; but considering the great thickness of this sequence, this range is not unusual.

The general geology, structure, and geomorphology of all the upper Missouri basins are similar, but their specific stratigraphy and structural histories differ in important respects. The smallest area that can be satisfactorily studied and described is a single basin, such as the entire Townsend and Prickly Pear Valleys; and Pardee's description of the Townsend Valley, based upon reconnaissance work, is a first and necessary step. Pardee, however, was unable to divide the Teritary beds into mappable units, but he did point out significant differences in their character and ages. In the present report, the Teritary beds of the Canyon Ferry quadrangle, are divided into five mappable units, but these units apply only within this quadrangle, and are doubtless only a partial section of such rocks in the Townsend and Prickly Pear Valleys. This fact is emphasized by the mapping of similar beds along the southwest flanks of the Spokane Hills as undifferentiated Tertiary. Other beds of Miocene age crop out just north and south of the quadrangle.

The westernmost and possibly the oldest of the Tertiary beds within the quadrangle consist of impure tuffs that flank the south and west sides of the Spokane Hills. All of them are mapped as undifferentiated Tertiary. Unit 1, of the Oligocene sequence, consisting of conglomerate, interbedded with red shale and some bentonitic beds, lies along the east side of the Spokane Hills. Unit 2, composed of volcanic conglomerate, tuffs, and other beds, extends eastward to the flats of the Missouri River. The principal deposits of bentonite occur at the base of this unit. Unit 3, composed mainly of reworked tuffaceous material without bentonite, extends from the Missouri River eastward, and occupies most of the foreland southwest of the Big Belt Mountains. Unit 4, consisting mainly of conglomerate with a calcareous matrix, lies still farther east, extending to the hard rock of the Big Belt Mountains. No unconformities or disconformities are known to separate units 1 to 3. On the contrary, all evidence indicates that these three units grade imperceptibly into one another; and the only basis for their separate delineation is the observable differences in their lithology. Unit 4, however, may lie unconformably on unit 3. Another group of Tertiary deposits, beyond the limits of the quadrangle, comprises certain slightly consolidated beds, which are exposed near the Winston bridge across the Missouri River and atop the Big Belt Mountains. The beds of this unit are probably of Miocene age.

UNDIFFERENTIATED TERTIARY BEDS

The undifferentiated Tertiary beds are poorly exposed in low bluffs along the north side of Beaver Creek, beginning 0.7 mile east of meridian 111°40′, and extending westward for about 2 miles. These beds also form the bedrock beneath low interstream ridges south and southwest of these bluffs. They extend southward, beyond the margin of the Canyon Ferry quadrangle, under the high gravel-covered ridge south of Beaver Creek, and northwestward under similar gravel-covered ridges in the valley of Spokane Creek.

The beds in the valley of Beaver Creek, are composed of soft, porous, slightly consolidated, buff sediments. A few grains as large as 2 millimeters are visible to the unaided eye. Under the microscope they are found to consist largely of angular grains and elongate shards of glass, as much as half a millimeter in size but averaging a tenth millimeter or less. Angular grains of quartz and plagioclase, together with rounded grains of country rock, iron ores, and rounded clusters of glass particles, constitute the remainder of these sediments. Other beds are composed dominantly of quartz and felspar, and subordinately of glass. Most of the igneous material had little or no transportation, but the rounded grains show an admixture of alluvial materials. These beds are classified as impure water-laid tuffs.

The attitude of the beds along the north side of Beaver Creek is indeterminate. West of the highway and north of Winston undifferentiated Tertiary beds strike N. 45° W., and dip 40° NE.; but farther north on the highway, near Clasoil, conglomeratic beds occur that are nearly horizontal, dipping slightly southwest. The absence of structural observations in the undifferentiated sequence close to unit 1 renders the relationship between these units indeterminate. It is possible that the undifferentiated Tertiary beds west and southwest of the Spokane Hills underlie those of unit 1; but lithologically they resemble most the beds of unit 3, and to a lesser degree the silty sediments interbedded with the Miocene gravels along the Missouri River just south of this quadrangle. The relative and absolute ages of the undifferentiated Tertiary beds along the west side of the Spokane Hills, and westward into the Prickly Pear Valley, are unknown, but they probably are a part of the Oligocene sequence.

OLIGOCENE BEDS

Unit 1.—The beds of unit 1 crop out in a belt along the east side of the Spokane Hills, for a distance of about 3 miles. At its southern end, this belt is overlapped on the west by terrace gravels; near its middle point, unit 1 is eroded so that the sediments of unit 2 are in contact with the Paleozoic rocks; but in the northern half of the belt, unit 1 rests on the hard rocks. The width of this belt near Beaver Creek is about two-thirds of a mile.

The boundaries of this unit are not well-defined. The beds include some conglomerate, which disintegrates by weathering so that the resulting material is not easy to distinguish from the terrace gravels.

Along its eastern boundary, the sediments grade into those of unit 2, so that this boundary is a matter of interpretation. The boundaries shown are therefore necessarily somewhat inexact.

The sediments of unit 1 are nowhere continuously exposed. A few outcroppings show that the unit consists of conglomerate and red shale, the latter including some bentonitic horizons, but the proportions of conglomerate to shale cannot be determined. At one site along the north side of Beaver Creek, the conglomerate is in place, and consists of well-rounded gravels, cobbles, and boulders, as much as 3 feet in diameter, with an average size of about 6 inches. They are mainly basic lava, with a small amount of quartzite. Some of the cobbles are exfoliated by weathering. The matrix consists of a yellow to reddish grit, made up of small well-rounded grains of limestone, and a small amount of other kinds of country rock. About a mile north of Beaver Creek, the red shale is exposed in an elbow-shaped bend of a small gulch. These beds are banded, tuffaceous, and bentonitic, and are composed mainly of grains as much as 0.1 millimeter in size, averaging perhaps 0.025 millimeter. The reddish-brown grains, originally glass, are altered to montmorillonite and kaolin. Angular grains of quartz, sericite, chloritized biotite, and altered iron ores are the other constituents. In the next gulch to the south, about three-fourths of a mile north of Beaver Creek, 10 feet of reddish-brown conglomerate and grit are exposed, the matrix of which is composed mainly of grains of limestone. At the north end of this belt, the conglomerate is similar in character, except that the cobbles and matrix are composed entirely of white limestone, derived from nearby hard rock in the Spokane Hills.

The materials of the sediments comprising unit 1 are primarily of local origin. Near Beaver Creek, the cobbles appear to be a mixture of rocks from the Elkhorn Mountains and the Spokane Hills. Farther north, the quartzite, limestone, and shale of the Quadrant formation were the source rocks; and the reddish color of some of them was transmitted to the rocks of unit 1. At the north end of the belt, the beds are clearly derived from white Paleozoic limestones.

Few observations of structural features are available, and therefore this belt has to be mapped largely on the basis of disintegrated conglomerate between which reddish shale debris and bentonite can be recognized. At the exposure on Beaver Creek, the beds strike N. 25° W., and dip 25° NE.; and at the site of the reddish brown conglomerate and grit, about three fourths of a mile to the north, the strike is N. 45° W., and the dip 15° to 20° NE. Assuming an average dip of 20°, the stratigraphic thickness of unit 1 is at most 1,200 feet; but an unconformable overlap on the east-sloping surface of the older rocks would diminish this figure greatly.

At a site north of Beaver Creek, unit 1 is in contact with, and dips under unit 2. This relationship, together with the data given above, show definitely that unit 1 underlies unit 2. But unit 1 has all the appearance of a local formation that was deposited along the slopes of the Spokane Hills, thus casting doubt upon the assumption that the nearby undifferentiated Tertiary beds are older than unit 1. The occurrence of bentonite, however, interbedded with the conglomerate of unit 1, shows that airborne volcanic debris was being, or had already been, deposited at or before the time when the conglomerate was deposited.

Unit 2.—The beds mapped as unit 2 extend from the eastern boundary of unit 1 to the last outcroppings of the Tertiary along the southwest side of the Missouri River. Farther south, however, beyond the limits of the quadrangle, beds that would be assigned to unit 3 crop out on the southwest side of the Missouri.

The lower third of unit 2 consists essentially of fragmental rocks of volcanic origin, which include about 25 percent bentonite, 25 percent partly bentonized tuff, 35 percent unaltered or little altered tuff, 10 percent tuffaceous conglomerate, breccia and sandstone, and 5 percent of other sediments including shale and chert. Except the bentonite, these beds range in color through shades of white, gray, greenish gray, cream, yellow, and buff. Most of them are well-indurated. The fabric is varied, ranging from fine-grained tuff whose grains cannot be recognized with the unaided eye, to grit and conglomerate containing pebbles that reach a maximum of several inches in diameter. The lithology is inconstant along the strike, so that parallel sections are difficult to correlate.

The conditions under which these sediments accumulated are revealed by more detailed examination. The pebbles of the conglomerates are of two kinds: cream-colored well-rounded pebbles of bentonized and kaolinized tuff, and dark-gray subangular to angular pebbles of glassy rhyolite that resemble smoky quartz. The soft bentonized pebbles are composed of rounded globules of glass, as much as 0.2 millimeter in size, aggregates of the same, glass altered to montmorillonite and opaloid material, and ghosts of twinned feldspar phenocrysts, set in a matrix of altered glass. The subangular to angular pebbles of rhyolite porphyry are holocrystalline, fine-grained and microporphyritic, and consist of microphenocrysts of sanidine and quartz in a fine matrix of the same two minerals, together with some biotite, and rarely sphene.

The tuffaceous grits and sandstones are cream-colored to greenish, the latter color indicating incomplete bentonization. The granules and coarse sand of these rocks are subangular to subrounded and comprise quartz, sanidine, microcline, oligoclase or andesine, and pieces of rhyolite; but there are also spherical to ellipsoidal ovoids of chert or opal, shards of the same, and porous glassy aggregates. The matrix cementing all these materials is fine glassy material, completely kaolinized or bentonized. Many of the finer-grained sandstones and tuffs are similar to the grits, except that some of them are undisturbed and unreworked airborne volcanic debris that settled directly in ponds. Such beds are commonly much whiter, and are little bentonized.

The most bentonized sediments, here called bentonites, consist largely of grains and aggregates of glass, altered to montmorillonite, sericite, kaolin, and other secondary products, together with grains of quartz, potash and lime-soda feldspar, and iron-ores in various stages of oxidation. Most of the glass, from which the bentonite was derived, is greatly altered. Some of the glass has an index of refraction between 1.50 and 1.51, but the bentonites are composite rocks, and it is probable that this glass is different from that from which the bentonite was derived.

The data presented above indicate that the lower third of unit 2 consists of two kinds of debris, first a glassy volcanic material that was subject to bentonization under certain conditions, and second a rhyolitic rock which is resistant to this process. The rounded grains of the bentonitic rock and the altered glassy aggregates show that the vulnerable glass accumulated and was to some degree consolidated, and probably largely bentonized under subaerial conditions, before it was transported by streams to its final resting place. The admixture of rhyolitic glass occurred before or during the stage when the bentonized glass was finally deposited.

Bentonites occur throughout this lower third of the sequence, but they are least impure and form the thickest beds, near the base. Along the strike these bentonitic horizons crop out to the northward for about a mile, and southeastward beyond the Canyon Ferry quadrangle for a mile and a half. At the surface these beds range through shades of green, yellow, and red, have a waxy appearance, and are invariably cracked as a result of swelling from the absorption of water and subsequent drying and shrinkage. Below the surface, where they have not been wet, they are gray, yellow, or brown compact rocks, that are readily cut with a knife. Their swelling factor is approximately two. Thin beds of chert, opalized in places, are commonly found at the bases of the larger beds of bentonite, though at many places they occur also within the bentonized zones.

The middle part of unit 2, according to Dreyer and Bush, consists largely of tuffs, about two-thirds of which are bentonized in various degrees. The beds of bentonite, however, are fewer and thinner, and aggregate probably 10 percent or less of this part of the sequence.

The proportion of chert, which appears to be a function of the degree of bentonization, also is much smaller. Tuffaceous conglomerate, grit, and sandstone continue as a part of the sequence.

The upper part of unit 2 differs in several respects from the two lower parts. First, it is made up largely of tuff, of which only about 10 percent shows any appreciable bentonization; and the amount of bentonite proper is negligibly small. Second, an appreciable part of the sequence, perhaps 10 percent, consists of a light-gray limestone, in part tuffaceous, which occurs in beds from 1 to 5 feet thick, increasing in thickness toward the top of the unit. Third, several beds of chocolate-colored sedimentary breccia consisting of volcanic detritus form a part of the sequence. These beds are composed of friable, pumiceous material, scoria, and ash, which form cliff-making beds that are useful as stratigraphic horizon markers. Finally, there occur at least two thin beds of snow-white diatomaceous earth in the upper part of the sequence. A specimen of this material was examined under the microscope, and 95 percent of it was found to consist of closely matted, unoriented skeletons of diatoms, about 0.02 by 0.003 millimeter in size. Several genera and a number of species are present. Small organisms, with circular cross sections about 0.006 millimeter in diameter, also are present.

The beds of unit 2, together with those of unit 1, compose a homoclinal sequence of rocks that dip east-northeastward at angles ranging from 15° to 30°. At the western limit of unit 2, the sequence is disturbed by several normal faults, the westernmost of which follows a fault plane striking north and dipping 70° E. One of the beds of bentonite apparently is displaced by this fault, but the throw appears not to be large. Other similar faults occur nearby to the east. The beds of unit 2, however, together with those of unit 1, appear to constitute a relatively undisturbed homoclinal sequence of rocks. The total thickness of unit 2 is given by Dreyer and Bush as approximately 1960 feet.

Unit 3.—The sediments comprising unit 3 extend from the Missouri River northeastward to an ill-defined contact with unit 4. This contact is definitely located in the valley of Cave Gulch, and is approximately located in the valley of Magpie Creek. Southeastward, however, it is inferred to lie between easily recognizable sediment of units 3 and 4. At the eastern end of the quadrangle, lack of exposures make it impossible to trace this boundary into the valley of Confederate Creek.

The sediments of unit 3, at their northwestern limit, have an irregular contact with the underlying rocks of Paleozoic and pre-Cambrian age. The southeastern limit of the belt is not known. The northeastern contact, with unit 4, trends east-southeast at the northwestern end of the belt, and southeast at the southeastern end. The south-western contact of unit 3 with unit 2 is unknown, but on plate 1 these sediments terminate where they are overlapped by the alluvial deposits of the Missouri River.

Exposures are too few throughout the belt to warrant a detailed stratigraphic section of unit 3, but the general lithology is fairly well known. Most of the sediments are cream to buff, porous, weakly consolidated, impure, water-laid tuffs, but they include a few beds of water-laid tuff that contain little or no admixture of nonvolcanic materials. Toward the top of the sequence, however, several beds of limestone occur, and above them the tuffs are interstratified with beds of grit and conglomerate, some of which have calcareous matrices.

The impure tuffs consist of varied amounts of colorless volcanic glass of all sizes and shapes, but principally angular and elongate shards and angular grains. They range in maximum size from 0.1 to 2.0 millimeter, and in average size from 0.1 to 0.005 millimeter, thus producing rocks that range from fine sandstone to siltstone. This glass, which constitutes a third to two-thirds of the impure tuffs, has an index of refraction of a little less than 1.52. The partly devitrified glass has a higher index of refraction. The other common constituents are angular to subangular grains of quartz, sanidine, orthoclase, with a little chloritized biotite. The more impure tuffs contain rounded grains of country rock, which include slate, chert, quartzite, and limestone, together with rounded grains of more basic plagioclase, epidote, muscovite, and iron ores. All these components, in various amounts, produce a considerable variety of rocks, here designated collectively as impure tuffs.

Beds of relatively pure water-laid tuff, ranging in thickness from a few inches up to 10 feet, occur here and there throughout the sequence. These beds are white or light yellow, and by their lighter color are in contrast with the impure tuffs. Such a bed, ranging in thickness from 4 to 7 feet, shows plainly in Magpie Bluff, east of the mouth of Magpie Creek. About 90 percent or more of such sediments consist of volcanic glass. These beds of tuff are lenticular, and cannot be used as horizon markers.

Tuffaceous sandstone and grit, with some beds of limestone, characterize the upper part of unit 3; and at or near the top of the sequence are several beds of conglomerate. The tuffaceous sandstone and grit differ in their degree of granularity from the impure tuffs, and also differ in having a larger proportion of quartz grains; some of them have calcareous matrices. The limestone is a well-indurated, white to cream-colored rock, that is characterized by numerous cavities, some of which are filled with secondary calcite. Small black veinlets of wad, as much as one-fourth inch in length, are common. The groundmass of calcite is resolvable only with a microscope. Five or six such beds of limestone, 2 to 3 feet thick, show in the lower valley of a small gulch about a mile northwest of Avalanche Creek. By an increase in the proportion of other minerals and fragments of country rock, these limestones grade into the calcareous grits. Beds of pure limestone are much less common than beds of impure limestone, or calcareous grit, leading to the belief that the pure limestone is lenticular and locally developed.

Noncalcareous grit and conglomerate occur near and at the top of unit 3. These are well-sorted rocks, composed of rounded pebbles and gravels, mainly quartzite and quartz, as much as 6 inches in diameter, in a sandy quartzose matrix. Conglomerate of this type occurs within 100 yards of the rocks of unit 4, in a little gulch tributary to Cave Gulch, about 3,500 feet east-northeast of the old site of Cavetown. Many beds of this type crop out in Cooper Gulch.

The beds comprising unit 3 dip from 10° to 30° northeastward, and therefore constitute a northeastward continuation of the homocline that characterizes units 1 and 2. In the vicinity of Magpie Creek, the width of the belt occupied by the beds of unit 3 is about 10,000 feet; and in the valley of the Missouri River, at the mouth of Beaver Creek, the width of the belt is probably about the same. The stratigraphic thickness is therefore on the order of 2,500 to 3,000 feet.

Unit 4.—The rocks of unit 4 are the easternmost of the entire Tertiary sequence. They are bounded on the southwest by the sediments of unit 3; and with numerous gaps, the contact between units 3 and 4 may be traced across the quadrangle. In the vicinity of Avalanche Creek, however, the distance of this contact from the southwestern face of the Big Belt Mountains is almost 4 miles, within which distance the Tertiary rocks are completely concealed by overlying gravels. The Tertiary rocks, moreover, do not crop out along the front of the mountains southeast of Magpie Creek. The true width of unit 4 in the central part of the quadrangle is therefore unknown.

At the northwestern end of the Tertiary belt, the rocks of unit 4 diminish in thickness to a thin wedge, and pinch out against the hardrock. The relations to the hardrock at the southeastern limit, however, cannot be determined, because the entire sequence close to the Big Belt Mountains is buried by the gravels that cover the foreland.

The rocks of unit 4 are dominantly conglomerate, with a minor proportion of gritty and sandy rocks that are similar to the matrix of the conglomerate. This conglomerate differs markedly from those that form the upper part of unit 3. It is brown to reddish brown, and consists of poorly sorted, subangular pebbles and gravels, which are composed largely of flat pieces of the rocks of the Belt series, notably the slabby material of the Empire and Spokane shales.

Moreover, the matrix is invariably calcareous, so that all these rocks are well-indurated and dense. The pebbles and gravels have a maximum size of approximately 6 inches, averaging perhaps 2 inches. At its extreme northwestern limit, where the unit wedges out, the rock is a porous white limestone, not greatly dissimilar to the nearby Paleozoic limestone. This unique part of the unit, which does not persist laterally, may be an ancient hot spring deposit.

Little is known of the structure of unit 4. The outcroppings are few where strikes and dips can be measured, but these observations, mainly in the valleys of Cave and Cooper Gulches and Magpie Creek, suggest a certain irregularity not present in the underlying beds of unit 3. Moreover, many of the beds appear to dip eastward and southeastward rather than northeastward. The beds, however, are close to the southwest front of the Big Belt Mountains, where post-Oligocene faulting is believed to have occurred. The irregular structure may, therefore, have been caused in part by such movements; but the lithology and structure of units 3 and 4, close to their contact, is sufficiently different to suggest that they are separated by an unconformity. These conditions, together with the 4 miles of gravel-covered pediment to the southeast, make an estimate of the total thickness impracticable. Any thickness from 500 to 5,000 feet is possible.

MIOCENE BEDS

Another group of Tertiary deposits, beyond the limits of the quadrangle, includes certain slightly consolidated beds that crop out along the route to Winston around the point of the spur between Confederate Creek and the bridge across the Missouri River. These beds are composed of well-rounded pebbles and cobbles, from 1 inch to 2 feet in diameter, averaging perhaps several inches. They range in thickness from a knife edge to 5 feet, and commonly lens out laterally within 50 to 100 feet. They are separated by beds of buff, slightly consolidated silt or clay, that resembles reworked tuffaceous material. Some of these beds are perceptibly tilted to the southeast, so that the thickness of exposed beds may aggregate 100 feet or more. The remains of a Miocene horse (*Merychippus* (?) *missouriensis*), described by Douglas (1908, p. 295), are reported to have been found in these beds. Pardee (1925, pp. 29–30) mentions several other localities, all south of the Canyon Ferry quadrangle where similar beds of Miocene age occur. They are generally softer than the Oligocene beds, are poorly exposed, and are less tilted. In the valley of Dry Gulch, about 6 miles southeast of Townsend, the discordance in dip between the older and younger beds is said by Pardee to be between 4° and 8°. Evidently these Miocene beds were laid down in shallow valleys carved in the older Tertiary beds, after the latter had been tilted to a third or a half of their present inclination.

Miocene gravels occur also in the extreme headwaters of Hellgate Gulch, at an altitude near 7,000 feet, and about 3,300 feet above the level of the Missouri River. At this place, beyond the limits of the quadrangle, the gently sloping hills form well-marked small parallel ridges, separated by undrained hollows. The tops of these hills consist of flows of basalt. This ridge and hollow topography is apparently the site of ancient gravels, which have moved downhill under the action of solifluction or some related process. These gravels are not exposed except at one place where they have been prospected for placer gold. The component pebbles, ranging in size to as much as 3 inches in diameter, are well rounded, but show secondary angularity resulting from disintegration. They are derived mainly from the Spokane, Greyson, and Newland formations. Several colors of gold were panned from a 60-pound sample of these gravels. These gravels are not necessarily contemporaneous with the Miocene deposits along the Missouri River; they may be considerably younger.

QUATERNARY SYSTEM

OLDER GRAVELS

The alluvial deposits are classified primarily on the basis of their relative ages, as determined by the altitudes of the bedrock underlying them. Approximate geologic ages have been assigned, but, lacking any paleontologic basis, they are merely an interpretation of the physiographic evidence. Four groups of alluvial deposits are thus recognized. The oldest consists of early Pleistocene gravels that cover the high hills on both sides of Beaver Creek, east of Winston. The second group comprises mid-Pleistocene gravels that mantle the Tertiary foreland, from the southwest side of the Big Belt Mountains nearly to the Missouri River. The third group consists of late Pleistocene gravels that lie on bedrock about 50 feet above the present level of the Missouri River. The youngest group includes all the gravel, sand and silt in the valley floors of the Missouri and its tributaries.

These designated groups do not constitute a complete record of alluviation in this part of the Missouri Valley. Gravels have been deposited on terraces that subsequently were completely destroyed; other terraces, with or without a mantle of gravels, are partly preserved and fit in between the four principal groups; and still other terraces and terrace gravels, at some distance from the Missouri, cannot be correlated exactly either with or between the principal groups, because ancient stream gradients cannot be determined without a topographic map, or many line traverses. The four groups herewith mentioned are therefore to be regarded merely as reference markers in the general history of alluviation.

Early Pleistocene gravels are found northeast, east, and southeast of Winston, on a high ridge that is really the southeastern extension of the Spokane Hills to the Missouri River. The valley west of this ridge, and east of the Elkhorn Mountains, was formerly either the lower course of Beaver Creek, or more probably that of a southeast-flowing master stream into which Beaver Creek discharged. In early Pleistocene time, however, Beaver Creek cut across the high gravel ridge, and began to carve its present course to the Missouri. At the northern limit of this ridge, where it forms the south end of the Spokane Hills, the high gravels lie upon the undifferentiated Tertiary beds and the beds of unit 1, but overlap onto the Paleozoic rocks. South of Beaver Creek, these gravels lie upon similar Tertiary beds that are exposed in gulches leading to the Missouri River. The old erosional surface atop the ridge is nowhere exposed, so that its altitude is not known; but the gravel-covered top of the ridge is about 300 feet above the level of the Missouri.

The character and thickness of the high gravels east of Winston cannot be completely determined, because they do not occur in exposed banks. Instead, they are found on top of a ridge, the fine material has settled, and the coarser material is disproportionately conspicuous. As seen in this environment, however, the gravels contain a high proportion of cobbles and boulders, which are as much as 5 feet in diameter, and are in places so prevalent as to form a boulder pavement. They are composed of volcanic rocks of intermediate and basic character, most of which were originally rounded to subrounded; but they are now spalled, exfoliated, and fractured, with the development of a secondary angularity. Some of the coarser-grained volcanics are deeply weathered. These gravels are obviously very old. A peculiar and unexplained feature along this ridge is the presence of conical or elongate piles of cobbles and boulders, up to 4 feet in height. One of the elongate piles is 15 feet long, 7 feet wide and 4 feet high. Some of the piles are isolated; others lie along straight lines for some distance.

The source of most of these high gravels was the Elkhorn Mountains, though local debris from the Spokane Hills is present north of Beaver Creek. The only east-flowing drainage of any size in this area, however, is that of Beaver Creek. But the gravel ridge is now separated from the Elkhorn Mountains by a broad valley which is 2 miles or more in width. The rounded character of the gravels suggest their original eluvial and alluvial origin, but their size and distance from their source, render doubtful the possibility that they could have been transported by Beaver Creek, or by the ancestral master stream that flowed in the valley between Spokane Hills and the Elkhorn Mountains. The suggestion is therefore advanced that glaciation existed in the Elkhorn Mountains in early Pleistocene time, and that these boulders were picked up by ice and transported to their present site.

Glaciation, if it existed in the Elkhorn Mountains, must have existed also at the same time in the Big Belt Mountains. No glacial gravels, or outwash deposits, however, have been recognized, but large erratics are present in some of the gravels that lie on top of the foreland between the Big Belt Mountains and the Missouri River, and in younger deposits close to the river. The erratics are large boulders of Flathead quartzite, up to 12 feet in diameter, that are believed to have been moved outward from the mountains by ice, and deposited originally on some ancestral surface that existed southwest of the Big Belt Mountains. They are now residual erratics that have been incorporated in younger gravels of several ages. The presence of such boulders on the low hills just east of the dam at Canyon Ferry suggests that ancient glaciation extended to the Missouri River.

The foreland that extends from the Big Belt Mountains nearly to the Missouri River is a gravel-covered pediment carved on the upturned edges of the Tertiary beds. Close to the river, as at Magpie Bluff, and other high bluffs to the southeast, the top of this foreland is about 150 feet above the level of the river, but rises gradually northeastward toward the Big Belt Mountains. At the surface of the foreland, the gravels are not everywhere visible, because they are covered with finer sediments, which in large part originated as coalescing alluvial fans on the pediment. Wind-borne material of later origin has still further obscured these deposits. Commonly, however, the gravels are revealed by shallow excavations.

The gravels of the foreland are well exposed at Magpie Bluff, east of the mouth of Magpie Creek, where they form a horizontal mantle overlying unconformably the tilted Tertiary sediments. The section here comprises 10 feet of loose gravels, underlain by 10 feet of gravels that are more or less cemented by calcite to a conglomeratic caliche. The upper 5 feet of the lower half of the section can be loosened and disintegrated with a pick, but the lower 5 feet has a hard calcified matrix. The cemented material consists of subangular to rounded detritus, ranging in diameter from a fraction of an inch to a foot; but at the top are some larger boulders, commonly quartzite, that reach a maximum of 2 feet in diameter. The overlying 10 feet of incoherent gravels are smaller and better rounded. Laterally this gravel deposit changes both in thickness and in the size of its detritus, but is everywhere calcified at its base. It is probable that some of the conglomerate in the uppermost part of Tertiary unit 3 has been reworked to form the lower part of the gravel deposit.

Near the outlets of some of the larger valleys the foreland is strewn with gravels, cobbles, and boulders, which constitute a younger set of mid-Pleistocene gravels. They were laid down in shallow valleys cut in the foreland, and represent the earliest gravels of the rejuvenation that lasted to the late Pleistocene and reduced the 200- to 150-foot base level of erosion to 50 feet above the present level of the Missouri. They are the oldest gravels that contain the erratics of Flathead quartzite mentioned above. Because they occur at, or nearly at the same level as the gravels on top of the pediment, and because they are imperfectly exposed, these ancient stream gravels are not readily separated from the gravel cover of the foreland. The best example of such gravels is on top the spur west of Cave Gulch, close to the mountains, where they have been exposed by mining operations. They are subangular gravels, composed largely of quartzite and subordinately of other kinds of country rock, mainly derived from the rocks of the Belt series. The average size is 1 to 2 feet in diameter, but they range upwards to large boulders, one of which was seen that measured 12 feet across. Close to the front of the mountains, these gravels are commonly limestone. The thickness is estimated to range from 10 to 20 feet. The tenor of these gravels in gold is said to have been high.

Stream gravels that lie at lower altitudes than those on top of Cave Hill, but at higher altitudes than the level 50 feet above the Missouri River, are present in the lower stretches of a number of the tributary valleys, and on top of some of the terraces below the 200- to 150-foot level. These deposits, similar in character to those on Cave Hill, occur along both walls of the lower valleys of Cave Gulch, Magpie Creek, and other streams to the east and west, where they were prospected and at some sites mined for their content of placer gold. Along the northwest wall of Magpie Creek, for example, these gravels, as exposed by mining, have a thickness as much as 15 or 20 feet, and consist of poorly sorted, angular to subangular cobbles and boulders, mainly quartzite, with a maximum size of 5 feet. The matrix is clayey, and layers of sand and silt are interstratified with the coarser beds. These deposits do not continue up the small gulches tributary to Magpie Creek, showing that they were deposited before these gulches were carved. They occur, and have been worked, at several levels on Magpie Creek and Cave Gulch.

The lowest, and youngest, of these Pleistocene bench gravels are well exposed by mining operations at the southwest end of the spur between Cave Gulch and Cooper Gulch, where they lie on tilted and beveled Teritary beds that form a bedrock surface 50 feet above the

Missouri River. The gravels lie in horizontal beds with an aggregate thickness of 35 to 50 feet, and are composed mainly of quartzite. The gravels in the lower beds of the section range from 1 inch to 12 inches in size, averaging perhaps 4 or 5 inches. They are separated by several beds of muddy sediment, up to a foot thick. The gravels in the upper part of the section are larger, up to 3 feet in diameter, and one boulder on the hillside, that comes from this deposit, is 8 feet in diameter. The tenor of gold is said to have been low; and evidently this statement was true, for a large volume of the deposit was never mined.

The youngest bench gravels at the mouth of Cave Gulch are of special interest. The Missouri River evidently flowed at one time at or nearly at the level of bedrock below these gravels. For causes yet obscure, the river at that time was aggrading its bed at the mouth of Cave Gulch and Magpie Creek, and this action helped to produce the considerable thickness of gravels at the mouths of these streams. After this aggradation, the bed of the river was about half a mile northeast of its present site, and at least 100 feet higher, so that it flowed close to the present site of the road along the northeast side of the river. As a result of this aggradation, the Missouri migrated southwestward, and was superposed on a spur of quartz monzonite, on that side of its valley. This aggradation and superposition are probably connected genetically with the back-hand drainage of the tributaries of the Missouri within this quadrangle, but a final answer to this problem must await geologic study over a much larger area.

YOUNGER GRAVELS

The alluvium that forms the present valley floor of the Missouri River and its tributaries consists of gravel, sand, and silt derived from many sources. Chief among these materials are the gravels derived from the hard rocks of the Big Belt Mountains and Spokane Hills, but much of this material has been deposited on terraces and reworked by later streams. The coarser debris lies mainly close to bedrock, and is not generally visible, except at places like the mouth of Magpie Creek, where these lower strata have been lifted to the surface by dredging. These gravels are well rounded and as much as two feet or more in size. Finer debris that acts as a matrix for the gravels, and also exists as strata of sand and mud, both between and above them, has been produced in part by comminution of the hard rocks, but in larger part by extensive erosion of the Tertiary lake beds. The thickness of these Recent alluvial deposits may be as much as 75 feet in the valley floor of the Missouri, but is much less in the tributary valleys, decreasing from the river to the face of the Big Belt Mountains.

IGNEOUS ROCKS

Intrusive igneous rocks of numerous kinds, and one group of lavas, are present in the Canyon Ferry quadrangle. The intrusive rocks include the Canyon Ferry stock, ranging in composition from quartz monzonite to granite; the lamprophyres and related rocks associated with this stock, likewise restricted to the Spokane Hills; the dioritic and gabbroic dikes and sills of the Big Belt Mountains; the andesitic dikes of the Big Belt Mountains and Spokane Hills; and the porphyritic quartz latite and associated intrusives of the Big Belt Mountains. The extrusives are basaltic lavas that occur mainly at or near the crest of the Big Belt Mountains; certain basaltic dikes are correlated with these lavas.

No extensive collections of these rocks were made. Instead, representative specimens were taken for the determination of their general petrographic character; and the many intrusives were correlated largely on the basis of field work. The following petrographic data indicate the main types of igneous rocks that were found.

MONZONITIC AND GRANITIC INTRUSIVES

Intrusives of quartz monzonite, granite and related rocks occur mainly at five localities. The Canyon Ferry stock, along the northeast side of the Spokane Hills, extends north-northwest for 4½ miles, and has a maximum width of 1½ miles. At its northern end, it is terminated by an incomplete sill or ring dike that follows around the north side of the hills. The elongation of the Canyon Ferry stock is generally parallel to the strike of the Paleozoic rocks that lie west of it, but locally the contact is irregular and discordant.

At the south end of the Spokane Hills is a small ovaloid stock, with a major dimension of 1½ miles from north to south, and a minor dimension of a little over a mile. About a mile to the north is a still smaller pluton, about half a mile long and wide, which is irregular in outline, and markedly discordant with regard to the surrounding country rock. A short distance northwest of the ovaloid stock is an elongate discordant pluton, or dike, that has a length of 2 miles, and a width ranging from 0.1 to 0.5 mile. This pluton transects Devonian, Cambrian and pre-Cambrian rocks.

A large body of quartz monzonite, which lies mainly outside the Canyon Ferry quadrangle, shows in the extreme southwestern corner. The size and shape of this pluton are not known. Its westward extension indicates that it is one of the outliers of the Boulder batholith.

The monzonitic and granitic intrusives, as mapped, include at least four types of igneous rocks. At the north end of the Canyon Ferry stock, as seen along the Canyon Ferry road, the intrusive is a coarse-grained granite. About a mile to the south, however, the granularity decreases and both granitic and monzonitic phases are present. Felsic and mafic segregates also have developed, so that aplitic dikes and larger bodies of lamprophyre are present.

The granitic intrusive at the north end of the Spokane Hills is a coarsely crystalline gray rock, mottled by light-colored crystals of quartz and feldspar, and dark-colored crystals of biotite and other mafic minerals. Some of the rock is notably porphyritic, with tabular crystals of feldspar as much as 25 millimeters or more in diameter. Under the microscope, the principal rock-forming minerals are found to be orthoclase and microcline, perthitic intergrowths of microcline and albite, plagioclase, quartz, and biotite. A little hornblende is present in some of these rocks, and rarely diopside. The common accessory minerals are apatite, iron ores, sphene, and zircon. Most of the plagioclase is oligoclase, but some of it is zonally grown, ranging in composition from oligoclase to labradorite. Quartz and orthoclase occur in graphic intergrowth but not commonly. Some of the plagioclase is partly sericitized, and epidote occurs as a secondary mineral replacing hornblende. Potash feldspar is so strongly dominant that these rocks are classified as granite or monzo-granite.

The intrusives that form the southern part of the Canyon Ferry stock, the smaller stocks farther south, and the pluton at the southwest corner of the quadrangle, consist of nonporphyritic rocks of medium granularity, most of which are darker in color, and more mafic in composition than the main stock. All these plutons are doubtless related genetically to the Boulder batholith, farther to the west. The minerals that compose these rocks are essentially the same as those in the main granitic intrusive, but the proportions are different. Most of the potash feldspar is orthoclase; the plagioclase, though zonally grown, is between oligoclase and andesine; and the ratios of potash to lime-soda feldspar range from 3:2 to 3:1. The former ratio is most common, so that most of these rocks are called quartz monzonite though some are granites. Both biotite and green hornblende are present, but one or the other of these two minerals predominates at different places. A little brown hornblende was seen in one specimen, and a little diopside, partly replaced by green hornblende, in another. The biotite is commonly somewhat chloritized. The chief accessory minerals are those named above, but west of this quadrangle, in the Boulder batholith, a site was found where the granitic rocks were sufficiently decomposed to be picked and panned. A sample of concentrates thus obtained by Mertie was examined by Prof. Adolph Knopf, of Yale University, and was found to contain, in addition to the accessory minerals cited, some grains of allanite and monazite.

FELSIC AND MAFIC DIFFERENTIATES

Both silicic and mafic differentiates are associated with the monzonitic rocks. The silicic fractions, which are aplite, are uncommon and occur as thin light-colored fine-grained dikes that cut the quartz monzonite and granite. At some places the boundaries of these dikes are indistinct; at others, the dikes are bounded by joint planes. One dike was found to consist mainly of orthoclase and quartz, graphically intergrown, with traces of chloritized biotite and iron ores. Such dikes are too scarce and too small to justify separate mapping.

The lamprophyric fractions of the monzonitic rocks crop out at two places on the Canyon Ferry road, one about half a mile northwest of Canyon Ferry, and the other about 75 feet west of the Flathead quartzite. The rock of the first locality has no distinct boundaries, and is included with the Canyon Ferry stock, but it may connect with the sill or incomplete ring dike that follows around the north side of the Spokane Hills. On its west side the ring dike is offset by faulting at several places, but probably was continuous originally with the lamprophyre at the second locality on the Canyon Ferry road. Southward from this site, the dike continues for several miles, and thereafter gives place to several larger bodies of similar intrusive rock. The southernmost of these bodies cuts sedimentary rocks as young as Devonian. These occurrences of the lamprophyres are shown on the geologic map, plate 1.

The lamprophyres are dark-colored rocks of medium granularity. They differ from the quartz monzonite in having a much higher percentage of dark-colored minerals, and in having a smaller proportion of quartz. The dark minerals, which commonly constitute two-thirds or more of these rocks, comprise hornblende and biotite; but diopside also is present in many of them, and is the predominant mafic mineral in a few of them. Iron ores are present, both as primary minerals, and as ordered clusters that suggest a secondary origin. Apatite is more common than in the quartz monzonite. Some of the quartz is free, but most of it is graphically intergrown with orthoclase. Plagioclase is the dominant feldspar, and is zonally grown, ranging in composition from oligoclase to labradorite. The ratio of potash to lime-soda feldspar varies, so that these intrusives range from quartz monzonite through granodiorite to quartz diorite; but the concentration of dark minerals renders lamprophyre a better generic designation for them.

The age of the monzonitic and related intrusives cannot be determined in the Canyon Ferry quadrangle from stratigraphic evidence. These intrusives are correlated however, with the Boulder batholith; and the latter, in the vicinity of Butte, intrudes lavas that are either of Upper Cretaceous or Eocene age. The Boulder batholith is therefore probably of early Tertiary age, and the monzonitic rocks of the Canyon Ferry quadrangle also are believed to have been emplaced in the Tertiary. Most of the lake beds in this area are of Oligocene age, but none of them has been invaded by these intrusives. Such negative evidence means little, but it tends to restrict the emplacement of the quartz monzonite and its satellitic rocks to the Eocene.

DIORITE AND GABBRO

The intrusives of diorite and gabbro group, so far as known, occur only in the Big Belt Mountains, within the several formations of the Belt series. Many of them have the appearance of sills, and some actually are sills; others are steeply pitching dikes that are approximately parallel to the bedding of the Belt series, but diverge at places to cross from one formation into another. The dike that is farthest northeast, is really a narrow elongate pluton, that is disrupted into segments by faulting. A general characteristic of all these intrusives is that they have been injected principally by pushing apart, or expansion, of the formations of the Belt series, without any perceptible assimilation or impregnation of these beds.

These intrusives are dark-greenish to greenish-black rocks of moderate granularity that weather brown. A few are porphyritic, showing phenocrysts of feldspars and mafic minerals, but this fabric is uncommon. Some of these rocks are mottled, owing to the presence of small areas of silicic minerals or of feldspar phenocrysts.

The characteristic modal minerals are plagioclase and augite, which commonly are intergrown in a poikilitic habit, such that the outlines of the pyroxene are determined by the crystals of feldspar. Iron ores are invariably present; and apatite, rarely also sphene, are the accessory minerals. Small amounts of quartz, or of quartz graphically intergrown with orthoclase (or microperthite), occur in the interstices between other rock-forming minerals, and constitute a characteristic feature. Primary green hornblende is a subordinate mafic mineral in some of these rocks, but in others this mineral occurs as a replacement of augite. Biotite and brown hornblende occur sporadically; and chlorite has developed from both the mica and the hornblende. Alteration of the feldspar has produced sericite, chlorite, and epidote. The plagioclase is zonally grown, but numerous determinations indicate that its average composition is either andesine or labradorite. This fact, together with the small but constant percentage of quartz and potash feldspar, indicates that most of these rocks are quartz pyroxene diorite or quartz gabbro. Some are diabasic in character.

Certain other dikes, of a more silicic type, have been identified in the general vicinity of the most northwesterly of these dioritic and gabbroic rocks, but they are either unmapped or have been included in this general group. One is a quartz monzonite, and another is a soda-syenite. The presence of these dikes suggest that intrusives of the Canyon Ferry stock also are present in the Big Belt Mountains, and their presence in the vicinity of the main diorite dike that is associated with gold mineralization, suggests that the Tertiary monzonitic and granitic rocks may be in fact the true source of the gold and copper mineralization of the Big Belt Mountains. The presence of granitic intrusives at the heads of Montana Gulch and Johnny Gulch also suggests the same conclusion.

No conclusive evidence exists regarding the age of this group of dioritic and gabbroic dikes. They resemble, to a degree, some of the lamprophyres associated with the Canyon Ferry stock, in the Spokane Hills; but the differences are so marked that the correlation of these two groups is considered improbable. The only stratigraphic evidence bearing upon their age is that they are restricted, so far as known, to the Belt series, and cut all formations up to the Empire shale. Therefore they are younger than most of the Belt rocks exposed in the Big Belt Mountains, but may not be younger than other Belt rocks that lie to the westward. Their minimum age is indeterminate. It is possible that they are of pre-Cambrian age, but they may be much younger. It is improbable that they are of Mesozoic or Tertiary age.

LATITIC INTRUSIVES

Latitic intrusives occur at two principal localities, both along or near the southwest face of the Big Belt Mountains. One is a dikelike pluton, with a maximum width of 1,500 feet, that extends from White Gulch to Bilk Gulch. The other, which lies at the head of the south fork of White Gulch, is a body of less regular shape that extends beyond the limits of the quadrangle. The two intrusives are closely related, and may connect below the surface. Dikes of quartz latite also are present, but not all of them were mapped. One occurs on Bilk Gulch about 6½ miles from its mouth. Much detrital debris of the same petrographic character lies on the foreland, between White Gulch and Dry Hollow. These intrusives, like the pyroxene diorite and gabbro, appear to have invaded the country rock by expansion of the beds along bedding planes.

The latitic intrusives are light-gray to medium-gray porphyritic rocks, consisting of phenocrysts of feldspar in a fine-grained groundmass. Judging from the detrital material found on the foreland, other facies of this intrusive also are present, which in their color, granularity, and composition, resemble granite and granite porphyry. The rocks found in the two intrusives bodies consist of phenocrysts of plagioclase, set in a matrix of orthoclase, plagioclase, quartz, and green hornblende, with smaller amounts of iron ores, apatite, and sphene. The plagioclase, particularly the phenocrysts, is zonally grown, and ranges in original composition from oligoclase to labradorite. Both are sericitized and albitized. Small amounts of diopside and biotite are present in some specimens, and one contains no quartz. Most of these rocks are considerably altered, with the development of chlorite, calcite, limonite, sericite, and albite. A considerable range in composition probably exists, as quartz latite, latite, dacite, and granite porphyry have been identified. The average type is best described as quartz latite porphyry.

These intrusives invade the Paleozoic rocks, but no additional stratigraphic data on their age are available. Possibly they are the finer-grained equivalents of the Canyon Ferry stock, and as such should be referred to the Tertiary.

ANDESITIC INTRUSIVES

The andesitic intrusives are mainly small dikes, but one larger intrusive body is known. Most of them occur in the rocks of the Belt series, in the valleys of Hellgate Gulch and Avalanche Creek, in Bilk Gulch, and on or along the divides between these streams. The single larger body crops out in the heads of Timber Gulch and Cow Gulch, whence it extends northwestward to and beyond Avalanche Creek, where it splits into two small dikes and disappears. It lies entirely within the Greyson shale; and to the southeast it extends beyond the limits of the quadrangle. This body and some of the dikes have been mapped. Some small dikes in the Big Belt Mountains, shown as andesite, are too greatly altered for accurate determination, and may not in fact belong in this group.

Somewhat similar, but possibly younger dikes of the same general character occur in the Spokane Hills south of the Canyon Ferry road and east of the lamprophyric intrusives. Another dike of the same general type was found on French Bar.

The rocks of this group in the Big Belt Mountains are dark gray, modified by shades of green and red. Most of them are either macro- or micro-porphyritic, the phenocrysts being invariably feldspar. The matrix is fine-grained. The feldspar, both of phenocrysts and groundmass, is plagioclase, which is entirely altered to sericite, chlorite, epidote, and albite, so that the original character is indeterminate. The general appearance, however, suggests a plagioclase of intermediate composition. The mafic minerals are completely altered to sericite, epidote, chlorite, and calcite, but certain prismatic and hexagonal outlines indicate that originally they may have been hornblende. A little quartz is rarely present, but most of it is probably secondary. Apatite and altered iron ores constitute the accessory minerals. The degree of alteration of these rocks suggests the designation of andesitic greenstone.

The andesitic dikes of the Spokane Hills are similar in general appearance to those described above, except that they are less altered and lack the shades of green and red. Some of them are porphyritic, others are not. The plagioclase is albitized and sericitized, but some unaltered biotite is recognizable. A little apatite and some iron ores also are present. These rocks, though mapped with the andesites, actually may be a fine-grained facies of the monzonitic lamprophyres. One of these andesites from French Bar is of special interest in that it was found to contain a few crystals of corundum. This is a very fine grained porphyritic rock, with small phenocrysts of feldspar and biotite. The groundmass is partly glassy. The plagioclase phenocrysts are altered to carbonates, but fine laths of unaltered plagioclase in the matrix are either andesine or oligoclase. Phenocrysts of corundum are very rare, and a little garnet is present.

The andesites of the Big Belt Mountains are restricted to the rocks of the Belt series. They are more altered than any other igneous rocks of the quadrangle. They may be of pre-Cambrian age, though no positive proof can be offered. The andesitic dikes of the Spokane Hills may be of Mesozoic or Tertiary age.

BASALT

Lava flows of basalt lie on the ancient surface that forms the crest of the Big Belt Mountains. The flows have been dissected by erosion so that isolated masses are separated from the principal flows. One such body occupies a rounded area about half a mile in diameter at the head of Grouse Gulch, a tributary of Magpie Creek. Basaltic lava as detrital material is prominent on the ridge northwest of Magpie Creek, just southwest of the southwest face of the Big Belt Mountains; but this material was not found in place.

Dikes of basaltic rock were observed at two places in the Spokane Hills, in the vicinity of the monzonitic lamprophyres, but were not separately mapped. It is probable that these dikes are genetically related to the lamprophyres.

Several types of extrusive basalt occur at the head of Grouse Gulch. They range in color from black to reddish brown, and in texture from massive to vesicular and amygdaloidal. Some specimens are holocrystalline and microporphyritic, with phenocrysts of labradorite in an intersertal groundmass of plagioclase and pyroxene, together with some iron ores, apatite and a little biotite. Other samples are partly glassy, with about the same mineral composition. The detrital material on the ridge northwest of lower Magpie Creek is nonporphyritic and partly glassy. The lavas are believed to be of Miocene age.

The intrusive basalt of the Spokane Hills is holocrystalline and microporphyritic, with phenocrysts of plagioclase, pyroxene, and olivine. It is classed as an olivine basalt.

STRUCTURE AND GENERAL RELATIONS

PRE-TERTIARY ROCKS

The Canyon Ferry quadrangle comprises three areas of structural unity: the rugged uplands of the Big Belt Mountains that are developed on steeply dipping and in part complexly folded and faulted beds of pre-Cambrian and Paleozoic age; the nearly isolated ridge that forms the Spokane Hills and is also developed on steeply dipping pre-Cambrian and Paleozoic rocks; and the lowlands, forming the broad valley of the Missouri River southeast of Canyon Ferry that are carved in the soft lake beds of Tertiary age.

BIG BELT MOUNTAINS

The Big Belt Mountains occupy the northeastern one-third of the quadrangle and extend beyond the limits of the map. Their structure is that of a broad, northwest-trending, uplifted arch. Along the central part of the arch, pre-Cambrian rocks have been thrust-faulted northeastward onto strata of Paleozoic age, as shown in the extreme northeast corner of the map; this faulting represents a stratigraphic displacement of several thousand feet. Southwestward the dips in the pre-Cambrian rocks are moderate, increasing in amount near the southwestern flank of the mountains. There the dips are generally nearly vertical or are overturned, and the overlying Paleozoic beds and in places the pre-Cambrian rocks are tightly folded and cut by high-angle thrust faults. The zones of faulting in the central part of the mountains and along the southwestern flank are extensions of regional deformation that continues northward into the Glacier National Park.

The southwestern front of the Big Belt Mountains is a zone of particular structural significance. This escarpment forms a broad, rather smooth curve that has a northerly trend in the southeastern part of the quadrangle, and a westerly trend in the northwestern part. The geologic structure along the southwestern front of the Big Belt Mountains is highly complex, and is characterized by an intricate pattern of folds and faults that affect the Paleozoic and pre-Cambrian rocks. The trend of these structures, however, is not the same as that of the southwestern face of the Big Belt Mountains, so that the structure obliquely transects the mountain front.

Tight folding along the southwestern flank of the Big Belt Mountains is best shown in the Paleozoic beds northwest of Hellgate Gulch and is nearly absent to the southeast. The beds dip steeply and in places are vertical or overturned, and the axial planes of these folds are vertical or dip steeply to the northeast. The folds are asymmetrical, as the beds on the north limbs of the anticlines dip at moderate angles and those on the south limbs dip steeply or are overturned. The crests and troughs of these folds are sharply flexed and the beds on any single limb have nearly constant dips, so that the folds in cross section resemble a series of tilted "W's." Where the folds are especially tight and well developed the Flathead quartzite, Meagher limestone, and Jefferson limestone are exposed in prominent ridges, whereas the shales and softer limestone beds are reduced in thickness and are poorly exposed.

The Madison limestone, which is restricted in the Big Belt Mountains to the western front of the range south from the vicinity of Little Hellgate Gulch, displays some structural features that emphasize and illustrate the structural complexity along the southwestern front of the mountains. In Hellgate canyon the Madison forms a great wall of rock standing nearly vertical but flattening higher up on the canyon walls to a normal westerly dip of 40°. Northwest along the strike, however, the Madison is progressively overturned, until near Little Hellgate Gulch it is rotated more than 180° and dips to the west at a low angle but in a reverse position. Northward from this site, the Madison is concealed below the surface of the pediment and cannot be traced. This structure is not well understood, but the available evidence suggests that a large recumbent fold is buried beneath the pediment to the southwest.

The high-angle thrust faulting in the Paleozoic rocks along the southwestern flank of the Big Belt Mountains is complex and difficult to analyze. It is best shown in the southeastern part of the quadrangle, where tight folding is negligible. Though some of the faults locally dip at low angles to the southwest, in general they dip steeply, mostly to the southwest but in places to the northeast. In the vicinity of White Gulch and in the area a short distance west of Hellgate Gulch the rocks are cut into slices, which are bounded by a series of branching faults. In the northwestern part of the quadrangle some of the tight folds pass laterally into strike faults that commonly dip steeply to the north or northeast.

High-angle normal faults of relatively small displacement, which trend generally northeast across the strike of the formations, also are present. The longest one, which is more than 4 miles in length, cuts across pre-Cambrian beds in the vicinity of Hellgate Gulch. Most of the others that were mapped are in Paleozoic beds, probably because these beds locally are more intensely deformed than are those of pre-Cambrian age, but possibly because faulting is more easily recognized in them. These faults, which cut and displace the folds, thrust faults, and igneous plutons, are unrelated to the small displacements that offset resistant formations, such as the Flathead quartzite.

The country southwest of the Big Belt Mountains is largely covered by Tertiary lake beds, so that few data are available regarding the structure of bedrock. In the northwestern part of the quadrangle, however, where the Tertiary beds are absent, a major structural break along the northwest continuation of the mountain front is more strongly suggested. There the Greyson and Spokane formations are apparently in fault contact with beds of Cambrian age, and the structural trend of the Spokane Hills abuts against that of the Big Belt Mountains. Furthermore a line of prospect openings, showing intensively brecciated rock and traces of ore minerals, is indicative of a zone of strong fracturing.

SPOKANE HILLS

The Spokane Hills form a rather broad north-trending ridge, developed on steeply upturned beds along the western limb of a general synclinal structure, the axis of which occupies the eastern slope of the Hills. This major structure is modified by minor folds and igneous intrusions.

Beds of Cambrian age everywhere form the highest parts of the ridge, dipping steeply in an easterly direction, though in places they are vertical or slightly overturned. They are flanked to the west by pre-Cambrian strata that stand nearly vertical. East of the Cambrian rocks progressively younger Paleozoic beds are present which have easterly dips that lessen eastward toward the axis of the syncline. Actually this synclinal structure consists of several discontinuous, slightly overlapping smaller synclines arranged in a staggered pattern, each associated with a complimentary anticline a short distance to the east. Not only are these folds somewhat irregular in form, but they are also partly interrupted by faulting and igneous intrusion. The northernmost syncline and anticline are rather open, nearly symmetrical, folds that plunge southward and abut against the Canyon Ferry stock. The folds southwest of Canyon Ferry are rather tight ones with steeply dipping limbs, but those in the southern part of the Hills are relatively open. Where the folds are rather tight the axes in places pass laterally into strike faults.

No true thrust faults were recognized in or along either side of the Spokane Hills, but steeply dipping normal faults that cross the strike of the formations at high angles are common. The longest one, which is in the southern part of the Hills, has a length of nearly 3 miles; but most of the others are less than a mile in length. In places the Flathead quartzite is cut by many small cross faults of small displacement that evidently extend only a short distance into the adjacent formations.

The north end of the Spokane Hills abuts structurally against the Big Belt Mountains. On the east side of the Hills for a distance of about 4 miles southward from Canyon Ferry the Paleozoic sediments are bordered on the east by the Canyon Ferry stock. Southeast, south, and southwest of the Hills the Tertiary lake beds lap onto the up-turned Paleozoic and pre-Cambrian strata, and the same relationship is present on the western and northwestern slopes of the Hills west of the quadrangle.

GENERAL INTERPRETATION

The general structure of the pre-Tertiary rocks within the Canyon Ferry quadrangle, ignoring faulting and minor folding, is that of a great synclinorium plunging southeast. But the Paleozoic and pre-Cambrian rocks of the Spokane Hills curve eastward and southeastward just west of the Canyon Ferry dam, suggesting that these rocks are the western limb of a smaller southeast-plunging syncline. On the other hand, this curvature might be regarded merely as a minor fluting in the major syclinorial structure. The Paleozoic and pre-Cambrian rocks of the Big Belt Mountains therefore may be interpreted either as the eastern limb of a syncline that is separated from the syncline of the Spokane Hills by an anticlinal structure, or as the eastern limb of the major synclinorium of which the structure west of Canyon Ferry is a mere fluting. The evidence in the valley of the Missouri River that might decide between these alternatives is completely concealed by Tertiary rocks, so that no decisive conclusion can be drawn; but the distance at the southeastern end of the quadrangle between the Paleozoic rocks of the Big Belt Mountains and those of the Spokane Hills suggests the possible presence of an intermediate anticlinal structure of major size, the axial plane of which plunges northwest. Such a structure would fit better with the presence of thrust faulting along the southwest flank of the Big Belt Mountains.

Strong evidence of thrust faulting along the south side of the Big Belt Mountains is found, but the direction and age of such faulting are not entirely certain. Pardee and Schrader (1933, pp. 24–25, 129–130) have described two thrust faults, called the Eldorado and Scout Camp overthrusts, that bound the Big Belt Mountains respectively on their southwest and northeast sides. Both of these fault planes are said to dip southwestward, so that the overthrust blocks have moved upward and to the northeast. Only the Eldorado overthrust is here considered, as the Scout Camp thrust lies mainly beyond the northeastern limit of this quadrangle. Most of the fault planes observed in the vicinity of White Gulch dip southwestward, but many of those farther to the northwest dip northeastward. These multiple fault planes, however, are best interpreted as fractures accessory to the main thrust faulting, so that none of them is the actual plane of thrusting. The actual plane of the Eldorado thrust is recognizable with assurance only in Clarks Gulch and Oregon Gulch, where it brings the Greyson shale into contact with the Flathead quartzite. Elsewhere to the east and southeast it probably passes beneath the Oligocene and later beds. Pardee considers that the Eldorado overthrust continues north-westward to connect with the Lewis overthrust, and southeastward to connect with the Lombard overthrust.

The intense folding and thrust faulting along the southwestern flanks of the Big Belt Mountains probably postdated the general fold-ing of the pre-Cambrian and Paleozoic rocks, but clearly antedated the faulting that produced the Tertiary lake basins and affected the Oligocene lake beds. The Lewis overthrust is considered by Pardee to be of Eocene age, and this interpretation fits with the evidence adduced in the Canyon Ferry quadrangle.

TERTIARY SEDIMENTS

The structure of the Tertiary beds is essentially that of a homocline dipping east-northeastward at angles from 10° to 30°, but the dips close to the Big Belt Mountains are more irregular and somewhat less than along the southwest side of the Missouri River. Thus, these beds overlap normally the eastern slopes of the Spokane Hills, in fact merging with these hill slopes; but along the front of the Big Belt Mountains they appear superficially to dip under the Paleozoic rocks. These facts and others presented below constitute evidence that a zone of normal faulting lies just south of the Paleozoic rocks that form the southern front of the Big Belt Mountains. This faulting is later than and unrelated to the thrust faulting that has affected the pre-Cambrian and Paleozoic rocks close to the southwestern face of the Big Belt Mountains. The trace or traces of this inferred faulting are largely covered by the sediments that lie atop the adjacent pediment; but where such traces might be discernible, they are obscured by the older thrust faulting heretofore mentioned.

The lithology, paleontologic data, and structure of the Tertiary rocks require further explanation. It is clear that a depositional basin was first provided in the early Tertiary; that the floor of this basin continued to subside during the deposition of the Oligocene sedi-ments; that these sediments were subsequently tilted by orogenic move-ments; and that much of this tilting occurred subsequent to sedi-mentation, though some of it took place before the youngest sediments were deposited. A depositional basin could have been produced either by downwarping or by faulting, but either gradual downwarping or intermittent normal faulting along the southwest side of the Big Belt Mountains throughout the deposition of the Oligocene beds would have resulted in a much greater range in the dips of these beds than actually exists. Therefore the basin of deposition is conceived to have been at the outset a graben, or block that dropped downward along two bounding faults, with little or no tilting. One of the bounding faults of such a structure must have been near, though probably southwest of, the present southwestern face of the Big Belt Mountains. The other bounding fault must have been some miles west of the Canyon Ferry quadrangle, possibly along the northwest side of the Prickly Pear Valley. At a much later date, after most of the sediments had been deposited in this basin, but probably just before the fourth Oligocene unit was formed, further movement began along the northeast side of the graben, that resulted in the east-northeastward tilting of the older Oligocene beds. This normal faulting, inferred in the preceding paragraph, continued after the youngest Oligocene beds were deposited; and judging by the similar tilting in Miocene beds farther south, these orogenic movements continued throughout most if not all of Miocene time.

The last disturbance that affected the Oligocene beds is normal faulting, striking north and dipping steeply east, which has affected mainly units 1 and 2. Similar faults, however, have been observed elsewhere in this sequence, even in the youngest Oligocene beds. These faults are not continuous for any distance, and their displacements are small, so that generally they have not been mapped. This faulting therefore constitutes the youngest orogenic movement of which there is any observable record.

The volcanism that supplied the debris of which the Oligocene beds are largely made must have originated during the Oligocene epoch, as the Miocene beds seem to be composed mainly of reworked volcanic detritus, mixed with coarser materials derived from the nearby hills. The Elkhorn Mountains, west of the Prickly Pear Valley, are composed of a variety of igneous rocks, including surficial volcanics of andesitic and more felsic character. Owing to their proximity and the character of the rocks composing them, these mountains, are regarded as the principal source of the ash and scoria found in the Oligocene beds. Such material could easily have been moved across the intervening distance of 30 to 40 miles by prevailing westerly winds.

ECONOMIC GEOLOGY

FIRST DISCOVERIES OF GOLD

The earliest reference to mineral wealth in Montana is said to have been made by Verendrye, in his report to the French Government in 1739. The Lewis and Clark expedition (May 1804–Sept. 1806) passed through the Canyon Ferry quadrangle on July 20–22, 1805, and gave the name of Whitehouse Creek to the stream later called Confederate Gulch, and still later Confederate Creek. Members of this expedition are said to have known or heard of the presence of gold in this region, but no true discoveries are recorded.

Two men have been credited with the discovery of gold in Montana. One was a Red River half-breed, named Francois Finlay (or Findley) and called Benetsee by his friends. In 1852 he discovered gold on Gold Creek, a small stream that heads in the northeastern part of Granite County, flows through Powell County, and joins Clarks Fork. The other discoverer was Samuel M. Caldwell, who is reported to have found gold in 1852 on Mill Creek, west of the Bitterroot River and about opposite Fort Owen. Most historians give priority to Finlay.

Ten years after the first discovery, the gold placer camps of Montana began to be established. Bannock City was founded with gold discoveries on Grasshopper Creek in 1862, Virginia City grew up with the discoveries along Alder Creek in 1863, Helena on Last Chance Creek and Diamond City on Confederate Gulch were established in 1864, and Pacific City on McClellan Gulch in 1865. By 1876, about 500 gold-bearing gulches had been found in Montana.

The original discovery on Last Chance Creek was made in July 1864 on a bar of Dry Gulch, one of its tributaries, not far from the present site of the Masonic Temple in Helena. The discoverers were a party of Georgians, of whom John Cowan, Robert Stanley, and Gabe Johnson were members. Not satisfied with their first discovery, they prospected northward as far as Sun River, but finding nothing better, they returned to Last Chance Creek in September, and began placer mining. The lateness of the season accounts for the name "Last Chance," which they gave to this stream.

This area was at first called the Rattlesnake district, but the name of Helena was soon adopted for the town. The overland distance from Helena to Fort Benton, head of navigation on the Missouri, was only 140 miles; and the town, supplied by this route, grew rapidly. A government assay office and substantial buildings were soon erected; and in 1874 the Territorial seat of government was transferred from Virginia City to Helena. From this settlement came those who discovered and developed the placers of the Big Belt Mountains.

The earliest discoveries of gold in the Big Belt Mountains are said to have been made by ex-Confederate soldiers. During the Civil War, Sterling Price's army of bushwackers and irregulars attacked the Union forces under Colonel Mulligan and after 7 days of fighting captured Lexington, Missouri. A short time later, however, Price's army was defeated and captured, and the Confederates were given the option of returning to their homes, or of banishment to the West. Many were thus transported up the Missouri to Fort Benton; and as the new discoveries of gold were being reported at that time, numbers of these men started for the gold diggings.

Along the southwest side of the Big Belt Mountains, gold placers were found between 1864 and 1866 in a number of streams, including Confederate Creek, White Gulch, Avalanche Creek, Hellgate Gulch, Magpie Creek, Cave Gulch, Clarks and Oregon Gulches, and York Gulch. Confederate Creek proved to be the most productive. On the northeast slopes of the Big Belt Mountains, less productive placers were located on Thompson, Indiana, Elk, Thomas, Benton, and Beaver Gulches, named from south to north.

The bonanza placers were worked out in a few years, after which the lower-grade placers began to be worked. At or about the same time, a search began for gold lodes, and one, the Old Amber lode, was discovered in 1870. Others were located in later years, but all of the higher-grade lodes were worked out by 1902. The lower-grade, or unoxidized ores, were worked for a number of years thereafter.

Both placer and lode mining of high-grade and medium-grade gold deposits are a thing of the past in the Big Belt Mountains. The mining of low-grade placers, however, has been carried on until very recent years, and the indications are that such work will continue on some stream and river bars in the years to come. The future of lode mining is doubtful. Only one small property is now being worked, and any similar mining of high-grade ores will necessarily be of small magnitude. In certain localities, however, as on the divide between Confederate Creek and White Gulch, it is possible that large-scale lode mining of low-grade ores some time may be attempted.

In the following paragraphs, the production of gold is given for a number of streams draining the southwest slopes of the Big Belt Mountains. Production data are given in dollars rather than in ounces of gold, because the original publication of the production was reported in this form. These dollar values should be increased by 69.3 percent, to accord with the devaluation of the gold dollar in 1934.

METALLIFEROUS DEPOSITS

LODES

Lode deposits in the quadrangle are valuable mainly for gold or copper; silver values are reported to be generally low, lead minerals occur at a few places, and one deposit containing manganese has been prospected. Exposures of rocks showing evidence of mineralization have been prospected at many places in the Big Belt Mountains and in the Spokane Hills, but few such deposits have been developed into mines, though a number of them have yielded test shipments of ore. None of the prospects in their present state of development, however, offers much, if any, encouragement for profitable exploitation.

GOLD LODES

MINERALIZATION

Most of the developed gold lodes of the Big Belt Mountains lie outside the Canyon Ferry quadrangle; none of them, except the Golden Messenger mine, has been worked for many years, so that their workings are no longer accessible. The mineralogy and economic geology of these deposits have been adequately described by Pardee (1933, p. 139–164). For these reasons, only the names, locations, and general character of these lodes will be stated, principally for the purpose of showing their relations to the gold placers.

A dike of dark-colored diorite extends from the east side of Avalanche Creek westward to and beyond Trout Creek. Partly as a result of faulting, and partly owing to erosion, this dike is not continuously exposed at the surface, but there is little doubt that the segments shown on plate 1 belonged originally to a single dike. Another intrusive of the same type extends from the east side of Avalanche Creek west-northwestward to and possibly beyond the northern limit of the quadrangle. Most of the gold lodes in the northern part of the quadrangle occur in or close to these intrusives, but the intrusives appear to be unlikely sources of the gold mineralization. Small dikes and bodies of a more silicic type, however, have been found near these larger dikes, thus suggesting that intrusives related to the Canyon Ferry stock may actually be the source of the mineralization, and that the dioritic dikes have functioned merely as channels for the mineralizing solutions. A small granitic pluton is present on the divide between Confederate Creek and White Gulch, at the heads of Montana, Greenhorn, Johnny, and Benton Gulches. No petrographic study of this intrusive has been made, but one specimen collected at the head of Montana Gulch is an alaskite; hence the conclusion that this is a granitic type of rock. A fifth significant intrusive is a light-colored quartz diorite, that was mapped by Pardee (1933, p. 134) along the northwest side of Confederate Creek, opposite the mouth of Boulder Creek.

These intrusives constitute the principal sites of gold mineralization along the southwest flanks of the Big Belt Mountains; and they in turn are the bedrock sources of the gold placers. This mineralization, however, is neither continuous nor of the same character or intensity at all places along these intrusives. One site of intense mineralization lies along or close to the principal dike of quartz diorite, in a stretch from the head of York Gulch to well beyond Trout Creek. In this area the Golden Messenger, Old Amber (Golden Cloud), Little Dandy and Daisy mines, together with a number of less well known gold lodes, were discovered and developed. Near the eastern limit of the dike, where it crosses Hellgate Gulch, both gold and copper mineralization is apparent, but the gold mineralization was weak, and no gold lodes have been developed in this area. Farther downstream on Hellgate Gulch, and also on Avalanche Creek, lodes containing ores of copper, lead, and zinc were found, some of which were developed to the producing stage. However, they are 2 to 4 miles south of the dike, and the genetic relationship is obscure. A third site of gold mineralization lies along and in the vicinity of the short dike of quartz diorite that crosses Avalanche Creek north of Thompson Gulch. Little lode development was done in this area, but the placer gold found in Avalanche Creek indicates clearly that gold lodes of some kind are, or at some time were, present in this zone of mineralization. A fourth locus of mineralization lies in and about the granitic intrusive at the head of Montana Gulch. Here the Miller mine was developed and, under the present name of the Champion Mine, is the only gold mine now being operated in the Big Belt Mountains. Scores of lode claims are held, and old prospect holes and small operating tunnels are everywhere evident. It is not know how many of these lodes were brought to the stage of production, but clearly this is an area of gold mineralization, that is an adequate source of most of the placer gold found in the valleys of Confederate Creek and White Gulch. The fifth site of mineralization was in the vicinity of the small dike on Confederate Creek, opposite the mouth of Boulder Creek. At this place, known as Norris Hill, much prospecting was done, and the Baker, Satellite, and other mines were developed. The gold from Norris Hill doubtless enriched the placers of Confederate Creek, but the real contribution cannot be gaged.

PRINCIPAL GOLD MINES

The largest and best known of the gold lodes in the Big Belt Mountains is the Golden Messenger lode, located on Browns Gulch (Dry Gulch) on the north side of Trout Creek, about 4,000 feet west of the forks of Kelley Gulch. The Golden Messenger mine lies just west of the northwest corner of the Canyon Ferry quadrangle. The history, development, and character of this property are given fully by Pardee (1933, pp. 146–157). The lode is a stockwork or quartz stringers and veins, some ranging in thickness from a knife edge up to 30 feet. They lie within the diorite, and strike generally north-northeast, with a westerly dip. The upper horizons of the lode were found to be oxidized ore containing free gold; but the lower zones are unoxidized, so that the gold-bearing sulfides had to be treated with cyanide. The recoverable tenor of the oxidized ore was $6 to $7 a ton, and the unoxidized ore had a tenor of $4 to $20 a ton. The lode was therefore of relatively low grade, differing in this respect from most of the other lodes in the area. The latest work on this property was done between 1926 and 1942, when the lode was intermittently mined. The mine was closed by order of the War Production Board in 1942.

Other lodes were located and developed in the vicinity of the Golden Messenger mine. Among them was the Little Dandy lode, located about half a mile east of the Golden Messenger, on the spur between Browns Gulch and Kelley Gulch. This lode consisted of a vein of quartz, from a few inches to 3 or 4 feet thick, that crossed the dike of quartz diorite, striking about north, and dipping 30° to 40° east. The unoxidized ore in the upper workings is reported to have had a high tenor in gold. Another lode of the same kind, called the Golden Charm lode, was located near the west end of the dike, west of the Golden Messenger mine. It consisted of several quartz veins ranging in thickness from one inch to a foot or more, in a country rock of shale. The ore of this lode also was of high grade. Both of these mines were practically worked out by 1900, but some work on the sulfide ores was later done at the Little Dandy mine.

The Old Amber, or Golden Cloud lode is located on the southeast side of York Gulch, a short distance downstream from the mouth of Rattlesnake Gulch. This was the first gold lode to be discovered and worked in the York district. The lode consisted of three or more quartz veins, striking about N. 70° W., and dipping 30° S., that lay south of the diorite dike. The ore in the upper levels was oxidized and of high grade; but that of the lower levels consisted of sulfide ores. Another property of the same general type, known as the Daisy lode, was located nearby, on the northwest side of York Gulch, north of the quartz diorite. Both of these lodes contained high-grade ores; both were developed by the same owners; and both were worked out by 1902.

The Champion mine (earlier called the Henry O. Miller mine) is located at the head of Montana Gulch, beyond the limits of the quadrangle. The ore body consists of seams of quartz and altered wall rock, in a metamorphosed shale close to the granitic intrusive. The ore is high grade, with a reported tenor of $100 a ton, or more. In the vicinity of this same intrusive are a number of other lodes, including the Schabert-Durant lodes at the head of Montana Gulch, the Hummingbird lode at the head of Johnny Gulch, and others. The upper or oxidized zones of all these properties were reported to have contained high-grade ore.

The Baker mine was located on Norris Hill, on the west side of Confederate Creek, opposite the mouth of Boulder Creek. This lode consisted of quartz veins in altered quartz diorite, ranging in thick-

ness from one to five feet, and striking northeast with a dip of 65°
to the northwest. The ore had a high tenor in gold. The Satellite,
and other veins of the same kind, also were located in this same
general vicinity.

COPPER AND OTHER LODES

No large base-metal lodes have been discovered or developed along
the southwest flanks of the Big Belt Mountains, but one copper property, the Argo mine, was operated for several years and produced
5,831 tons of ore, with an average tenor of 25.7 percent copper. Pardee
(1933, pp. 164–171) describes or mentions about a dozen other metalliferous lodes, none of which ever had any significant production.

The Argo mine is located on the east side of Hellgate Gulch, about
3 miles upstream from the place where that stream issues from the
Big Belt Mountains, and about 8,000 feet upstream from the north
side of the Flathead quartzite. Details regarding the discovery,
development, and operation of this property are given fully in Pardee's
report. The Argo ore body is located at or near the contact between
the Spokane and Greyson shales, and strikes about N. 70° E., dipping
steeply north. In the oxidized zone, which extends downward to a
depth of 100 feet, the ore body is 4 to 5 feet wide, and consists of
spongelike brown iron oxides and malachite, together with a little
unoxidized chalcopyrite intergrown with quartz. Copper-stained
quartz stringers lead into the wall rock. Both red and green argillite
constitute the country rock, but it is recorded that the ore occurs
only in the green argillite. Below the zone of oxidation, the vein
ranges in thickness from a few inches to 18 inches, and consists of
ankerite, quartz, and chalcopyrite. Some of this ore had a tenor
of 26 percent copper. Ores of lower tenor were concentrated to this
grade before shipment. The vein is cut and displaced by faulting.
The ore body was developed for 500 feet along its strike, and for
600 feet down its pitch. The principal development of this mine
occurred during the First World War, when the price of copper was
high.

Small pods or discontinuous stringers of vein quartz crop out in
many places in the pre-Cambrian rocks of the Big Belt Mountains.
Many have been tested with shallow prospect pits, particularly in the
vicinity of the Argo and Old Amber mines and along the diorite dike
north of Thompson Creek. Few if any of these quartz veins have
visual evidence of metallization, and none of these prospects is located
on the map.

From the mouth of Cave Gulch to the west edge of the map the
contact between the Paleozoic rocks to the north and the Tertiary
and pre-Cambrian rocks to the south marks a zone of major faulting

and brecciation. Numerous pits, trenches, small shafts, and adits explore traces of ore minerals that have been deposited along this zone. Near Cave Gulch and for a quarter of a mile to the west scattered nodules and stringers of manganese oxides have been found in soft brown gouge. Along the same zone and just north of the small monzonite stock that is centered between Oregon Gulch and Clarks Gulch several small shafts open a zone of brecciated Greyson shale partly impregnated with copper carbonates. West of Clarks Gulch a series of moderately extensive adits explore a large low-dipping gouge and fracture zone. A very few small stringers of galena ore were found in all of these workings.

Pardee describes or lists the following properties, which are tabulated herewith:

1. The Conshohocken group of claims, on the west side of Hellgate Gulch, a short distance upstream from the Argo mine. This is a copper property, and the ore minerals mentioned are quartz, ankerite, limonite, chalcopyrite, and chalcocite.

2. Mike Finch and Hellgate claims, east and down the slope from Conshohocken group. These are copper claims.

3. Claims of Ideal Mining Co. on west side of Hellgate Gulch, opposite the Argo mine. This is a copper property, whose ore contained quartz, ankerite, and chalcopyrite.

4. Rex claim, on Gabisch Gulch. This is a copper claim, showing ores of quartz, ankerite, and chalcopyrite.

5. Whitmire claims, on northwest side of Magpie Creek, a short distance downstream from the mouth of Coxie Gulch. This is a copper lode. One surface sample of 16 tons showed 17 percent copper.

6. Sibyl Ann claims, in Coxie Gulch. Ore consists of quartz, ankerite, and chalcopyrite.

7. Big Copper lode, located 800 feet north of the Golden Messenger gold lode. The ore is mainly ankerite, with some other minerals and a very small amount of copper-bearing ores.

8. Copper Queen Co. lode, on ridge south of York Gulch, about three-fourths of a mile west of Old Amber gold lode. Ore minerals are mainly calcite and quartz, but include a little chalcopyrite.

9. Lode on Upper Number 2 Gulch, tributary of White Gulch. Ore is quartz, ankerite, and a little chalcopyrite.

10. Korizek mine, on south slope of French Bar. Ore consists of quartz, barite, and galena, with smaller amounts of pyrite, chalocopyrite, cerussite, malachite, and chrysocolla. The mineral cuprodecloizite, containing vanadium, copper, zinc, and lead, is also present. An ore pile is reported to have assayed from $60 to $97 a ton in silver, lead, and copper.

11. Ankerite lode, on south side of Missouri River, opposite French Bar.

12. Tom lode, half a mile north of ankerite lode. The ore consists of iron oxides, quartz, cerussite, and galena.

13. Manganese lode, alonge south face of limestone, where Cave Gulch emerges from the Big Belt Mountains. The ore consists of manganese oxides, derived probably from rhodonite and rhodochrosite.

In addition to these properties, the following also should be mentioned:

1. The Doolittle copper mine, on northeast wall of Doolittle Gulch, about 300 yards from Avalanche Creek. The ore consists of quartz, barite, chalcocite, bornite, malachite, and other sulfides. One sample is reported to have assayed 7 percent copper and 6 ounces in silver. Some ore was shipped from this property.

2. Another copper lode, similar to the Doolittle lode, on Kellogg Gulch, a tributary of Avalanche Creek.

3. Prospect at mouth of small gulch on west side of Hellgate Gulch, about 2⅓ miles upstream from Argo Mine. Ore consists of stringers of chalcopyrite and galena, in the quartz diorite dike that crosses here.

4. Lead- and copper-bearing veins associated with a dioritic dike have been prospected at two places along Hellgate Gulch about 2½ miles upstream from the Argo mine and also just east of the divide between Hellgate Gulch and Spring Gulch. The veins follow the contact of the dike with the Newland limestone or fractures within the dike, and they range from a few inches to a foot or so in width. The veins are composed of quartz containing some pyrite, galena, copper sulfides, and perhaps sphalerite.

5. Narrow copper-bearing veins have been prospected on the west side of Avalanche Creek about a half mile south of Kellogg Gulch. The veins range from 1 inch to 6 inches in width and are weak and discontinuous. They are partly filled with barite and carbonate gangue, through which small masses of chalcopyrite and secondary copper minerals are irregularly distributed.

6. In the Spokane Hills, as in the Big Belt Mountains, many outcrops showing some evidence of mineralization have been tested with shallow prospect pits or with a limited amount of underground work. In the pre-Cambrian strata along the west side of the Hills quartz pods or stringers are common along fractures or at the contact with igneous rocks, whereas in the Paleozoic beds mineralized fractures and contacts are indicated by silicified limestone or jasperoid. Traces to small amounts of secondary copper minerals are present at many places, and in some of the prospects a few grains of copper, lead, or iron sulfides are exposed.

PLACERS

EARLY PLACER MINING

CONFEDERATE CREEK

The placers of Confederate Creek, though outside the Canyon Ferry quadrangle, deserve mention because the lower part of this valley lies within the quadrangle, and may contain gold placers. Confederate Creek heads in the Big Belt Mountains and flows southwestward for about 12½ miles to the Missouri River. Cement Gulch, one of its tributaries from the northwest, joins Confederate Creek about 11 miles from the Missouri. At distances of 2, 2⅓, and 2¾ miles downstream from Cement Gulch, three other tributaries enter from the northwest, known respectively as Montana, Greenhorn, and Dry Gulches; and about 0.3 mile downstream from Dry Gulch, Boulder Creek enters from the east.

According to several accounts, the discoverers of gold on Confederate Creek were Jack Thompson, Washington Baker, and Pomp Dennis. Ascending the Missouri River in the summer of 1864, this party camped at or near the mouth of Confederate Creek, where Thompson sank a prospect hole, and found gold-bearing gravel valued at 10 cents to the pan. The tenor of the gravel increased as they prospected up the valley. A second discovery farther upstream was made a few weeks later by John Wells and another small party of propectors. The bonanza gravel on Confederate Creek was located on December 3, 1864, at a point on the creek somewhere near the mouth of Dry Gulch. This gravel was reported to have had a tenor of $180 to the pan.

Later discoveries in Confederate Creek were made by a party from Virginia City, known as "the Germans," including Charles Fredericks, John Schonneman, and a number of others. These men found rich placers farther upstream on Confederate Creek, and also in Montana and Cement Gulches. Eventually Confederate Creek was mined from Cement Gulch downstream for 7 miles. Some of the German group discovered very high grade placers along a northwest terrace of Confederate Creek, between Montana and Greenhorn Gulches, which was called Montana Bar. A downstream continuation of Montana Bar, known as Diamond Bar, was later located west of Greenhorn Gulch; and still later a high-level placer, known as Boulder Bar, was discovered on the opposite side of Confederate Creek, on the spur north of Boulder Creek.

The discovery of the rich placers of Confederate Creek and its tributaries, and of Montana and Diamond Bars, led to a sudden and tremendous boom. The town of Diamond City was built on the terrace between Greenhorn Gulch and Dry Gulch, and its population quickly reached 5,000 people. Stores and hotels were built, and for 3 or 4 years, the town was very prosperous. Thereafter it gradually declined and although most of the town was destroyed when Diamond Bar was mined, it still boasted 3 stores and 2 hotels in 1875. By 1880, however, the population had decreased to 64, and by 1895 little was left of this settlement.

Montana Bar was one of the highest grade placers on record. Discovered in October 1865, it was opened up and worked out from May 9 to August 15 of 1866, producing gold valued at nearly $1,000,000. Later reworking of the gravels and additional mining operations along the margins of the pay streak and at higher levels on the bar, yielded an additional $500,000 to $800,000. The bedrock level of Montana Bar was about 50 feet above the level of Confederate Creek; its length in an easterly direction was about 1,600 feet; its greatest width was 400 feet; and its total area was about 5 acres. The richest part of the pay streak, however, was much narrower, as it was completely covered by claims 200 feet wide, and is said to have been contained in an area of about 2 acres. The thickness of the deposit ranged from zero at the wall of the valley to 40 feet along its north side. The gold was coarse, and it is recorded that one nugget valued at $900 was recovered. Very rich streaks and crevices were found on or in bedrock, and from one such place a pan of gravel valued at $1,000 is said to have been taken. The average tenor, however, gaged by the recovery from 2 acres was at least $10 to the square foot of bedrock; and if the average thickness of the deposit is taken as 18 feet, the tenor was not less than $15 to the cubic yard. The fineness of the gold is not known to the writer, but one shipment of gold, reported to have weighed somewhat less than 2 tons was valued at $900,000. The fineness cannot be determined from these figures, because it is not known how much black sand the shipment included. If the net weight of the shipment is estimated at 3,800 pounds, the fineness in gold was not less than 785; but if this weight included much black sand, the fineness was considerably higher.

Diamond Bar, west of Greenhorn Gulch, is the downstream continuation of Montana Bar, and has about the same level above the creek. It extended from Greenhorn Gulch intermittently to Dry Gulch, so that the length of the pay streak in an easterly direction was about 2,300 feet. The average width of the pay streak was reported to be about 250 feet. The tenor of the gold-bearing gravels on Diamond Bar was less than on Montana Bar, but the deposit was nevertheless high grade.

Boulder Bar, on the south side of Confederate Creek, north of Boulder Creek, is about 150 feet higher than the valley floor. The gravels of this bar were of high grade, and the deposit is now mined out.

Nearly all of the gold on Confederate Creek was transported into its valley via Cement, Montana, and Greenhorn Gulches, principally from the first two. No placers were found upstream from the mouth of Cement Gulch. Johnny Gulch, a tributary of White Gulch, which heads against Montana and Greenhorn Gulches, also had high-grade placers. And Benton Gulch, one of the placer-bearing streams in the northeast side of the Big Belt Mountains, likewise heads in this highland area. It is therefore evident that the bedrock sources of most of this gold were lodes localized at the heads of these streams, on the high dome known as Miller Mountain, lying beyond the limits of the quadrangle.

Cement Gulch was mined for a distance of 1.4 miles airline from its mouth, and according to Pardee (1933, pp. 172–173), two claims near its mouth, aggregating 400 feet in length, yielded $400,000 in gold. Pardee estimates the total yield of Cement Gulch as between $800,000 and $4,000,000. Montana Gulch was similarly mined for 1¼ miles above its mouth, and produced between $600,000 and $3,000,000 in gold. The total production from Confederate Creek and its tributaries is estimated by Pardee as approximately $12,000,000.

The complete history of placer mining in the valley of Confederate Creek has not been recorded. Doubtless much of the ground was first worked by drift mining, which required the construction of long underground bedrock drains. Later an organization known as the Flume Co. hydraulicked the stream gravels from about a mile downstream from Boulder Creek to half a mile upstream from Montana Gulch. The pay streak in this stretch is said to have been from 40 to 150 feet wide, and the gravels 30 to 40 feet thick. The tenor of the gravel in these operations is unknown but was probably high grade.

A home-built dragline scraper was constructed and put into operation in 1940 by a man named Woodward, who in this and the two following years worked about a half mile of marginal ground that had been left by the Flume Co. Woodward worked a width of 20 to 30 feet, but he was handicapped by old tailings, and did not get to bedrock. The tenor of the gravel which he handled in these operations is reported to have been 25 cents to $1.00 per cubic yard. Another dragline scraper plant is reported to have worked on Confederate Creek during 1940 and 1941.

A summary of the recent placer mining on Confederate Creek has been given by Lyden (1948, p. 17), who states that Boulder Bar was worked with a power shovel and a stationary washing plant from 1938 to 1941, inclusive. The plant probably cleaned at least 400,000 cubic yards of gravel, the tenor of which was 23 cents or more in gold to the cubic yard. As no sluice water was available at the level of Boulder Bar, the washing plant was located in the valley floor. A part, or all, of this later mining was done by a man named Lematti; the deposit is now owned by Charles Sheridan of Helena.

Farther down the valley of Confederate Creek, but within the hills, a stretch of three-quarters of a mile was worked from 1939 to 1941, by a man named Burkstrand. The work was done by a dragline dredge, often called a "doodle-bug" plant, which consists of a dragline excavator, feeding gravel to a floating washing plant similar to a dredge but without a digging ladder. The gravel is reported to have been 32 feet thick. The pay streak here is about 200 feet wide, and as a result of drilling was reported to have a tenor of $1.20 a cubic yard; but the excavator did not reach to bedrock, and only a part of the gold therefore was recovered.

WHITE GULCH

The valley of White Gulch is shown on the Canyon Ferry map, plate 1, from the mouth of Number 16 Gulch downstream to its mouth. Between Number 16 Gulch and the southwest face of the Big Belt Mountains, the valley floor is narrow, and the stream gradient is steep owing to lack of adjustment of the upper valley to the present base level of the Missouri River.

About 2 miles upstream from Number 16 Gulch, Spring Gulch enters from the northwest, and Miller Gulch enters from the southeast a few hundred yards downstream from Spring Gulch. Johnny Gulch, a tributary from the southeast, joins White Gulch about 3 miles upstream from Number 16 Gulch; and Park Gulch, from the northwest, enters about a hundred yards farther downstream. Johnny Gulch heads against Montana and Greenhorn Gulches, tributaries of Confederate Creek, in the high dome-shaped ridge called Miller Mountain.

The history of placer mining on White Gulch is not completely known. Apparently the original discovery was made on Whites Bar, a low terrace on the northwest side of the creek, which extends from the mouth of Park Gulch downstream for 0.6 mile to the next gulch, and continues intermittently downstream for 0.7 mile to Spring Gulch. Gold was discovered on Whites Bar on May 2, 1865, and shortly thereafter White City was located on this bar, about ¼ mile downstream from Park Gulch. In the early boom days, the population of this town was 1,000 persons.

The surface of bedrock on Whites Bar, at the old site of White City is about 30 feet above the level of the creek, but slopes gently upward for 150 yards to a gravel bluff about 40 feet high which marks the limit of the old hydraulic mining operations. At the base of this bluff, but several feet above bedrock, the writer took four pans which showed considerable gold. Evidently mining terminated at this point, not because the lateral limit of the pay streak had been reached, but because the recovery in gold no longer paid for removal of the thick overburden. All of Whites Bar but only parts of its downstream continuation to Spring Gulch were worked. Just downstream from Spring Gulch, along the northwest wall of the valley, other workings show on a small terrace called Frazer Bar, which is about 75 feet above the level of the creek. Similarly, just downstream from Miller Gulch, on the southeast wall of the valley, is another small terrace called Ryans Bar, where many good-sized nuggets, some weighing as much as 3 ounces, were found.

The floor of White Gulch was mined by drifting from some undetermined point, possibly as far downstream as Number 16 Gulch, upstream to Johnny Gulch; and shallow drift workings and open-cuts are visible in Johnny Gulch for at least a mile above its mouth. Placer workings are present in White Gulch for only a few hundred yards above the mouth of Johnny Gulch. It is evident that most of the gold found in White Gulch was transported from the lodes on Miller Mountain via Johnny Gulch, though some gold may have entered the valley by way of Miller Gulch. Pardee (1933, p. 179) estimates that the total production of gold from White Gulch had a value between $1,000,000 and $1,500,000.

AVALANCHE CREEK

Avalanche Creek is shown on plate 1 for a distance of about 6 miles upstream from its mouth. At this point a south-flowing tributary called Shannon Gulch enters; and about 500 feet downstream another tributary, known as Cayuse Creek, enters from the east. Upstream from these two gulches are a number of others which are named. The two most important are Nary Time Gulch, entering from the northeast, and Thompson Gulch, entering from the west, about 2.1 and 2.7 miles upstream respectively from the mouth of Shannon Gulch. The valley floor of Avalanche Creek, for about a mile upstream from the southwest face of the Big Belt Mountains, has a steep overall gradient, showing incomplete adjustment to the latest base level of the Missouri River.

Few placer mining data on Avalanche Creek are extant. Some prospecting may have been done in the stretch between Doolittle Gulch and Shannon Gulch, but no signs of any extensive mining remain. From Shannon Gulch upstream for some distance, Tom Shannon did considerable drift mining about 1900; and from Needham Gulch upstream to the mouth of Thompson Gulch the valley floor was drifted almost continuously. This work was done by Barnes and Huncks, Frank Carney, Art Spencer, and others, about 1900; but more recent work in this stretch has been done by John H. Bird.

The placer workings also continue up Nary Time Gulch for some distance. Thompson Gulch was worked by hydraulic mining. No placer mining in the valley floor of Avalanche Creek was done upstream from Thompson Gulch, but a low terrace along the southwest wall of the valley has been worked at a number of sites throughout a distance of a mile. This terrace is about 50 feet above the level of the creek.

Some of the gravels underlying the gravel floor, as at the mouth of Shannon Gulch, had a thickness of from 20 to 35 feet. Drift mining of such ground required underground bedrock drains of considerable length. To pay for the amount of necessary dead work, these gravels must have had a fair tenor in gold, though they are known not to have been as high grade as those on White Gulch. The value of this gold, on the basis of a sale made in 1939, was $28 an ounce, thus indicating an approximate fineness of 800. The pay streak is narrow, but it is possible nevertheless that sometime this gravel may be reworked.

HELLGATE GULCH

Hellgate Gulch was doubtless prospected at or about the time when mining was in progress in adjacent valleys to the southeast and northwest, but evidently no high-grade placers were found, as no tailing piles of any size are visible. Some drift mining, however, is known to have been done years ago.

In the headwaters of Hellgate Gulch, beyond the limits of the quadrangle, at an altitude of approximately 7,000 feet, and almost on the divide between Hellgate Gulch, Magpie and Avalanche Creeks, gravels were found that are capped by basaltic lava of Tertiary age. These gravels had been prospected at several sites and three pans taken by the writer from the prospect dumps were found to contain several fine colors of gold. These are the oldest auriferous gravels known in the Canyon Ferry quadrangle.

MAGPIE CREEK

Magpie Creek heads in the Big Belt Mountains, flows southwesterly and discharges into the lower end of Lake Sewell. The extreme headwaters of Magpie Creek lie outside the Canyon Ferry quadrangle, but all of its valley is shown wherein placer mining was done.

A highland known as Hedges Mountain lies at the head of Bar Gulch, a tributary of Magpie Creek from its northwest side; and Cave, Oregon, Kingsbury, and York Gulches, farther to the west, likewise head in this highland area. Hedges Mountain, and particularly its southern and southwestern slopes, is the site of the bedrock sources of the placers in all of these streams. The pay streak in the valley of Magpie Creek therefore starts in Bar Gulch, and from its mouth continues downstream.

Magpie Creek, like White and Hellgate Gulches and Avalanche Creeks, flows through a constricted stretch in its valley, near the southwest flanks of the Big Belt Mountains. The valley floor within this stretch has a gradient higher than it has farther upstream or downstream, and represents, as in the other valleys mentioned, an incomplete adjustment to a base level lower than that which formerly controlled the upper valley. Magpie Creek differs, however, from White Gulch and Avalanche Creek, in that gold placers occur both above and below the constricted part of its valley. Gold placers are present from the mouth of Bar Gulch downstream for about 2 miles; and beginning beyond the southwest face of the Big Belt Mountains, continue intermittenly for about 2 miles to its mouth.

Little is known of the history of mining on Magpie Creek. Gold was discovered in this valley probably about 1870, but apparently little or no mining was done until the eighties. The valley floor of Bar Gulch was worked for a distance of 1¼ miles from its mouth, mainly by drift mining. The old shafts appear to have been about 15 feet deep, indicating that thickness for the deposit. The gravel is small, and angular to subangular. Magpie Creek, from the mouth of Bar Gulch downstream for 2 miles was worked continuously by drifting in the eighties by Courtney Sheriff, and later by the Magpie Mining Co. and individual operators. In the early drifting operations by Sheriff an underground bedrock drain was constructed, so that a stretch of half a mile, the length of the drain, was never worked. According to Pardee (1933, p. 178), the Magpie Mining Co. sank a shaft in this stretch in 1928, and discovered that the depth to bedrock was in excess of 60 feet. The pay streak in the part of the valley that was mined is said to have had a maximum width of 150 feet; and the gold, which sold commercially at $17.25 an ounce, is reported to have had a fineness of 825. A 2-ounce nugget was discovered by Sheriff. The production in this part of the valley of Magpie Creek and from Bar Gulch, is estimated by Pardee at $180,000.

Gold placers were worked at a number of sites along Magpie Creek, for a distance of 2 miles above its mouth. In the upper part of this stretch the valley floor was not mined, but four cuts were worked along the northwest side of the valley floor, at distances of 300 to 30 yards from the creek. The largest and most northeasterly cut was about 1½ miles from the mouth of Magpie Creek, and was 1,500 feet long with a maximum width of 200 feet. The second cut, which begins 1,900 feet downstream from the lower end of the preceding one, has a length of 1,250 feet and a maximum width of only 100 feet; but farther upon the valley wall, two smaller cuts were worked at a higher level. The northeast end of the lower cut, curves laterally into a small gulch; and this same feature is noticeable in a third cut, which is really a downstream continuation of the second, interrupted by the small gulch up which it follows. The fourth or most southwesterly cut is close to the mouth of Magpie Creek, is 1,000 feet long and about 175 feet wide.

The southwest wall of Magpie Creek, from the face of the mountains downstream for a mile, is a rather steep gravel bank, cut by numerous small gulches with a maximum length of a quarter of a mile. Small-scale placer mining has been done in five of these gulches, and between two of them. Southeast of this valley wall is a well-developed terrace that forms the divide between Magpie Creek and the headwaters of Little Hellgate Gulch. The top of this spur, and its corresponding level in the northwest side of Magpie Creek, are parts of terrace 2, as heretofore described. Old prospect pits and small opencut workings are visible on this spur, southeast of Magpie Creek, but the corresponding terrace northwest of Magpie Creek is not gravel covered. The distribution suggests that the course of Magpie Creek, at the time when it debouched from the hills onto terrace 2, was southsouthwest, conforming with its course just within the Big Belt Mountains; and this in turn indicates that its backhand drainage across the Tertiary foreland to the Missouri has not been modified.

The gravels forming the walls of Magpie Creek, and covering the terrace on its southeast side, are angular to subangular, poorly sorted, and unconsolidated or only very slightly so. They include coarse cobbles and boulders, the latter ranging in size to a maximum of 5 feet in diameter. Much clayey material is present in the matrix, and sandy or clayey strata are commonly interbedded with the gravels. All these gravels, and their included placers, are ancient, probably Pleistocene, detritus from the Big Belt Mountains, which was deposited when terrace 2 was formed, and has in various degrees been reworked by streams of later date.

The Recent gravels in the valley floor of Magpie Creek have been dredged from its mouth upstream for nearly a mile. This work was done during 20½ months of the years 1911–13, and is reported not to have been a profitable enterprise. The dredge, built by the Union Iron Works, had buckets with a capacity of 5 cubic feet, and was driven by electric power. Pardee (1933, p. 178) estimates that the gold produced by this dredging operation had a value of $50,000.

CAVE GULCH

Cave Gulch resembles White and Hellgate Gulches and Avalanche and Magpie Creeks in that if flows through a constricted stretch in its valley, just inside the Big Belt Mountains, where its gradient is abnormally high. Cave Gulch resembles Magpie Creek in that it has two pay streaks, one above and one below the constricted stretch in its valley.

A small settlement, called Cavetown was located in Cave Gulch, at the mouth of a small tributary from the east, about 0.7 mile from the Missouri River. The settlement of Canyon Ferry, however, founded in 1865, must have been the principal supply point for this area. No trace of Cavetown remains. The lower pay streak on Cave Gulch was discovered by Dan McIntosh in 1866, before the upper one had been found.

Cave Gulch differs from Magpie Creek in that its course across the Tertiary foreland is south-southwest instead of southwest, thus conforming with its general course through the Big Belt Mountains. Unlike Magpie Creek, gravels are distributed on the high terrace on both sides of its lower valley, though most of the placer mining was done along its west side. Numerous small mining and prospecting pits, however, show along its east side, particularly along the headwater stretch of the small tributary of Cave Gulch, that enters at the old site of Cavetown.

Two types of placer deposits were worked along the west side of the lower valley of Cave Gulch. Both are terrace deposits, but they lie at different levels. Near the mouth of Cave Gulch, and extending northwestward to Cooper Gulch, the spur separating these two drainage channels is wide and spatulate, and consists of thick deposits of gravel which extend northeastward for about a third of a mile. These gravels have been mined, and therefore their stratigraphic section is well exposed in the old mining faces. They range in thickness from 25 to 50 feet, and rest on the eroded surface of Tertiary beds, which dip gently east-northeast. This old surface of erosion lies about 50 feet above the present level of the Missouri River. The gravels are poorly sorted, subangular, and range in size from 1 inch to 5 feet, averaging perhaps 3 inches. Large boulders, ranging in size from 2 to 5 feet in diameter, are rare, but are invariably quartzite, cleary derived from the Flathead quartzite. Among the finer gravels is much gray, green, and red slate, eroded from the several formations of the Belt series. Thin layers of mud, laterally discontinuous, and ranging in thickness from 2 to 10 inches, are present throughout the section, but are much more prevalent in its lower part.

These gravels were worked by hydraulic methods along the lower west wall of Cave Gulch, along the front facing the Missouri, and most extensively in a long cut between Cave Gulch and Cooper Gulch. The gold is said to have been distributed throughout the gravels, though most of it was close to bedrock. Panning near bedrock by the writer resulted in the recovery of several fine colors of gold. It is reported that mining operations were not profitable, probably because too much gravel had to be handled for the amount of gold recovered. This report is doubtless true, because obviously a very large volume of unworked gravel remains.

The spur separating Cave Gulch and Cooper Gulch rises gradually to the northeast, and becomes narrow along its summit; and about 0.6 mile from its southwestern end, it rises abruptly to a higher level, estimated to be about 175 feet above the valley floor of Cave Gulch. This higher part of the spur, known as Cave Hill, is everywhere covered by gravel, which has been worked almost continuously for a mile southwestward from the face of the Big Belt Mountains. These gravels are much thinner than at the southwest end of the spur, and in the stretch from the rise of the spur to the face of the mountains are estimated to have a thickness of from 10 to 20 feet. They are composed largely of quartzite, and have an average size between 1 and 2 feet. Large boulders of quartzite, eroded from the Flathead quartzite, are common. One had a diameter of 12 feet.

The desposits along the spur described above were worked by small-scale hydraulic mining and by open-cut shoveling-in operations. The gravels are said to have had a high tenor in gold, estimated by Pardee (1933, p. 178) to have been $5 to the square yard of bedrock. Pardee gives also the probable production in gold as $500,000.

Little is known of the character of the upper pay streak in Cave Gulch, within the Big Belt Mountains. The placers lay mainly in the valley floor, and are said to have been worked by drift mining for a distance of 2 miles. Pardee estimates that the gold produced in this upper stretch of the valley had a value of $400,000. It is probable that this estimate is too high, as most of the gold from Cave Gulch is said to have come from its lower valley.

COOPER GULCH

Cooper Gulch is a small western tributary of Cave Gulch which joins the latter close to the Missouri River. This gulch, with a total length of a mile, heads in the Tertiary foreland, and taps no bedrock that could be a source of gold. The gold of Cooper Gulch came therefore from the old valley of Cave Gulch, and the placers were formed by a reworking of the terrace gravels on Cave Hill and on spurs leading westward therefrom. These placers were mined first by Van Camp and Boswell, soon after the deposits on Cave Hill were found; and later by Boswell and Tandy.

The principal placers of Cooper Gulch are those along the southwest end of the spur which lies between this gulch and Cave Gulch.

These gravels were worked by hydraulic mining, but like those at the mouth of Cave Gulch, are said to have had a tenor too low for profitable mining. The tops of several small spurs in the valley of Cooper Gulch are likewise covered by gravel, which has been reworked and reconcentrated as a pay streak in the valley floor. It is reported that this pay streak was worked over a distance of 2,000 feet, and numerous old pits and storage reservoirs for water tend to corroborate this report. Cooper Gulch is totally devoid of water except during heavy rains, and the water for this placer mining is said to have been brought by ditches from larger valleys to the east.

OREGON GULCH AND CLARKS GULCH

Oregon Gulch, with an airline length of 5½ miles, heads in Hedges Mountain, and flows southwesterly to the Missouri River. About three-quarters of a mile from its mouth, Clarks Gulch enters from the west. Oregon Gulch and Clarks Gulch, like others farther east, are not in adjustment to the present base level of the Missouri, and therefore have constricted stretches where rejuvenation is working upstream. The stretch of oversteepened gradient on Oregon Gulch extends from a point one-fourth of a mile below the mouth of Holiday Gulch downstream for a mile to the mouth of Clarks Gulch; and a similar stretch is present on Clarks Gulch.

The placers of Oregon Gulch were discovered in 1865 or 1866, and were largely worked out by 1870. The original site of discovery was about 1,000 feet downstream from the mouth of Holiday Gulch. From this point, a pay streak in the valley floor was worked intermittently to the head of the valley. According to William H. De Borde, who was born and raised at Jimtown, the paystreak in the valley floor extended from the upper end of the gorge upstream for 2,000 feet. Little work was done in the next 2,000 feet; but thereafter the pay streak continued for another 1,000 feet. Most of this was drift mining. In addition, terrace gravels along the northwest wall of Oregon Gulch have been mined by open-cutting, starting about a quarter of a mile above the gorge and continuing intermittently upstream for about half a mile.

Clarks Gulch has a sinuous valley about 3 miles in length, measured along the course of the creek. Upstream from the outcrop of the Flathead quartzite, the stream gravels of Clarks Gulch and its tributaries were worked to their extreme headwaters. Below the gorge, the terrace gravels of Clarks Gulch, including Centennial Bar on the west side of the valley, were worked to its mouth, and for some distance down the valley of Oregon Gulch.

Jimtown was built at the head of Clarks Gulch, and served as a supply center for Oregon Gulch and Clarks Gulch. Water for mining these gulches was taken from the drainage basin of Trout Creek, and was carried to the head of Clarks Gulch by an 8-mile ditch, which took several years to construct. The gold recovered from Oregon Gulch and Clarks Gulch is said by Pardee (1933, p. 177) to have had values, respectively, of $500,000 and $300,000.

YORK GULCH

York Gulch heads against Cave Gulch, and certain gulches that flow northwest to Trout Creek, and drains west-southwest for 5 miles to join Trout Creek. The valley is deep and narrow. The principal tributary is Kingsbury Gulch, which enters from the southeast, about 1.2 miles above the mouth of York Gulch. A smaller tributary, called Rattlesnake Gulch, joins York Gulch from the south side about 1¾ miles upstream from Kingsbury Gulch. The Old Amber, one of the best known gold-quartz lodes in this area, is situated on the southeast wall of the valley of York Gulch, about midway between Kingsbury and Rattlesnake Gulches.

York Gulch is next in importance to Confederate Creek, as a producer of placer gold in the Big Belt Mountain district. Gold is said to have been discovered in the lower valley of York Gulch by a man named Price, in 1864, at the time of the Sun River stampede, but organized placer mining apparently began in the spring of 1866. In 1867 the town of New York, later called York, was established at its present site at the junction of York Gulch and Trout Creek, north of the former. Another town, now abandoned, know as Brooklyn, was established south of York Gulch. A large mining population occupied this district, and in the late sixties York was next to Diamond City in size.

The gold placers of York Gulch were mined from its mouth upstream to Rattlesnake Gulch. Placer mining also was done on Kingsbury and Rattlesnake Gulches. The greatest amount of gold was recovered from the stream gravels reaching from Kingsbury Gulch to Trout Creek. In this stretch, the pay streak appears to have ranged from 50 to 150 feet in width, averaging 100 feet or more, though a much smaller width was worked in the original drift mining. The thickness of the gravels is about 40 feet. Most of the gold was found on or close to bedrock, but two runs of gold, of different grades, are said to have been present. At the outset, this part of the pay streak, together with the stream gravels upstream from Kingsbury Gulch, yielded richly; but by 1874 all the gravel that could be worked by drifting had been exhausted. Thereafter the ownership of all the claims downstream from Kingsbury Gulch was consolidated in the Trout Creek Mining Co.; and from 1888 to 1891, Frank D. Spratt, principal owner of the company, worked this stretch by hydraulic methods, producing the open-cut that is now visible. Water for these operations was obtained from the McCune ditch, which carried 600 miners inches for 3½ miles from Trout Creek.

The stream gravels of York Gulch, from Kingsbury Gulch upstream to Rattlesnake Gulch, are 10 to 20 feet thick, and constituted a pay streak that ranged in width from 15 to 50 feet. This gravel was mined by drifting but was never hydraulicked owing doubtless to lack of water. The valley floor in this stretch is narrow, and has a fairly steep gradient, so that successive floods have swept away most of the old tailings. Little evidence therefore remains of the early drift mining.

Rattlesnake Gulch, for 500 feet upstream from its junction with York Gulch, is a steep gorge with a very high gradient, in which no placers were found. For 3,000 feet farther upstream, the valley floor was evidently worked intermittently by drifting; and at this point, a gulch entering from the southeast shows some old gravel piles. York Gulch, above the mouth of Rattlesnake Gulch, was apparently not worked. Kingsbury Gulch, the principal tributary of York Gulch is said to have had no continuous pay streak, because the gold was erratically distributed, or "spotted." The valley floor of Kingsbury Gulch, however, was worked intermittently to the headwaters, though some of this work was not profitable.

York Gulch, until 1870, is said to have produced $800,000 in gold, obtained entirely from underground mining. The total production, including all mining of every kind on the gulch and its tributaries is not definitely known, but is reported to have been as great as $5,000,-000. Even if the production was only half as much as this, York Gulch rates next to Confederate Creek as a producer of placer gold.

TROUT CREEK

The valley floor of Trout Creek, for a mile and a half downstream from York has a maximum width of 300 feet, but from this point to the Missouri River it flows in a narrow gorge. Two shafts have been sunk in the valley floor, one a half mile and the other two miles below York. From these shafts it is known that the thickness of the gravels in this stretch ranges from 60 to 90 feet. In the early days of mining in this district, the average flow of water in Trout Creek is said to have been 2,000 miners inches, but it is much less now.

The long pay streak in York Gulch, extending from some of its headwaters downstream to its mouth, naturally led to the belief that workable gold placers should be present in the valley of Trout Creek, below York. About 1890, Augustus N. Spratt started an open bedrock drain at the mouth of Trout Creek, which was intended to reach bedrock at some point well below York. On the completion of this drain, he planned to mine the gravels of Trout Creek by hydraulic methods, but the project never materialized to the stage of mining. The shafts mentioned above were sunk for the purpose of drift mining, but the flow of water was too great to be controlled. The gravels below York were drilled in 1931, and were undrilled a few years ago, but the results of this exploration are not known. If the valley floor of Trout Creek is ever mined, the work will be done with a dredge, or perhaps with a dragline excavator, and a pumping system to provide sluice-water under pressure.

MISSOURI RIVER TERRACES

A number of terraces covered by gold-bearing gravels are present along the Missouri River, from Canyon Ferry downstream for 20 miles or more, measured airline along the river. French Bar, the terrace farthest upstream, is situated on the southwest side of the river, about 2½ miles downstream from the Canyon Ferry dam. Only the eastern part of French Bar lies within the Canyon Ferry quadrangle. In the early days of placer mining in this district, a settlement called French Bar existed for some years at that site, close to the river.

The other principal bars are Gruel Bar, on the northeast side of the river, about 4½ miles northwest of Canyon Ferry; Spokane Bar, on the southwest side of the river, opposite Gruel Bar; McCune (McQueen) Bar, on the east side of the river, about 6 miles northwest of Canyon Ferry; and Eldorado Bar, on the north side of the river, about 7¾ miles northwest of Canyon Ferry. Other bars to which reference is made in the literature are Ruby Bar, believed to lie between Gruel and Eldorado Bars, Danas Bar, on the peninsula east of the lower end of Lake Helena; and American and Mings Bars, along the east side of the river, much farther downstream.

Gold placers were found on all these bars soon after the discovery of gold in the nearby streams. On French Bar, according to Pardee (1933, p. 180), the principal bedrock terraces lie at altitudes of 200, 240, and 260 feet above the level of Hauser Lake which, opposite French Bar, is about 15 feet above the original level of the river. Mining was done, however, at numerous other levels. The gravels lying on these terraces range in thickness from 1 to 6 feet, above which lies a varied thickness of sand, silt, and alluvial material. The old mining pits range in width from 50 to 400 feet, and extend eastward for over a mile. One assay of the gold recovered from French Bar in 1927 shows that the melted bullion contains 864.25 parts gold, 135 parts silver, and only 0.75 part dross. This amount of dross is unusually small. Water for these mining operations was obtained from two long ditches which took water from the headwaters of Spokane and Beaver Creeks, and carried it northwestward along the southwest side of the Spokane Hills.

The area worked on French Bar is estimated by Pardee to have been 150,000 square yards, and the tenor about $10 a square yard, resulting in a gold production of $1,500,000. Another estimate given to the writer, however, was 50 cents a cubic yard, which if the total overburden was as much as 18 feet, would indicate a value of $3.00 to the square yard of bedrock. Possibly some of the bonanza gravel was as rich as stated by Pardee, but probably the average tenor was much lower and the total production not over $500,000.

Spokane Bar is 3½ miles downstream from French Bar, on the same side of the river. According to Pardee, the gravels on Spokane Bar lie about 100 feet above the old valley floor of the river, and are composed of well-rounded cobbles, 6 to 8 inches in size, that are partly indurated by a rusty-red cement. The gold-bearing gravels are only a few feet thick, and are overlain by 5 to 20 feet of barren gravel, sand and soil. The old workings are about 3,000 feet long, 50 to 500 feet wide, and from 6 to 25 feet deep. The value of the gravels is stated by Pardee to have been possibly $3.00 to the square yard of bedrock, thus indicating a possible production of $500,000. It is know, however, that the tenor on the river terraces decreased downstream. If the average tenor is taken as 40 cents a square yard of bedrock the average thickness of the gravels as 18 feet, and the average width of the paystreak as 400 feet, the production would have been only $120,000.

Little information is available on the other terraces, between Spokane Bar and Eldorado Bar, but recent dredging on Eldorado Bar has yielded some dependable data. At this site two well-marked old erosion levels exist, one about 160 feet above the level of Hauser Lake, and the other about 35 feet below the level of the lake, though minor intermediate terraces also are present. Allowing a fall of 5 feet to the mile, from Canyon Ferry to Eldorado Bar, these levels would be 220 and 25 feet above the original bed of the river. At the upper level, where the latest work was done, the face of the cut shows 8 feet of gravel, overlain by 15 feet of fine hillside wash. The gravels range in size to a maximum of 30 inches, with 18-inch boulders fairly common, though the average size is about 5 inches. The overlying debris is almost entirely derived from the Belt series, and ranges in size from one-eighth to three-fourths of an inch.

The gold recovered by the dredge is said to have occurred as small grains, with no nuggets of any size. The fineness of the gold ranged from 870 to 910; and 29 assays, supplied to the writer by Mr. Owen Perry, showed an average fineness of 890.2 in gold and 105.6 in silver.

The remainder was dross. Some platinum metals also were recovered every year, and one estimate indicates the presence of 7.86 ounces of platinum, 0.56 ounce of osmium-iridium, and 0.42 ounce of palladium to the 100 pounds of well-cleaned concentrates. Ruthenium and rhodium, though probably present, were not reported by Wilberg Bros., the analyst.

In the early days of placer mining, Eldorado Bar was worked by hydraulic mining. Water for this purpose was taken from Trout Creek and was carried from 15 to 16 miles in a ditch that delivered 1,000 miners inches. In December 1938, a dredge was built on Eldorado Bar by the Perry-Schroeder Mining Co. and was operated until June 1944. This dredge was constructed of steel pontoons, and the hull measured 100 by 50 feet. It worked with two spuds, aided by bow and stern lines. The stacker was 100 feet long. The digging ladder consisted of 93 buckets of 6 cu. ft. capacity, and the gravels were cleaned in a 40-foot trommel screen, perforated with holes ranging from three-eighths to three-fourths of an inch in size. Four flow lines led from the trommel, with two 42-inch jigs in each line. There were also 16 cleaner jigs. The dredge was equipped to dig to a depth of 48 feet, and had to do so on the lower pay streak, where it worked below the level of Hauser Lake.

This dredge worked both the lower and the upper pay streaks. In the 6½ years during which it was operated, it handled about ten million cubic yards of material. The value per cubic yard is not known, but allowing for the known decrease in tenor downstream from French Bar, and for the fact that some of this ground had been worked by hydraulic methods in the early days, the average tenor for this yardage is estimated to have been between 15 and 25 cents to the cubic yard. The byproduct output of platinum metals was small, amounting probably to about 25 ounces a year, or roughly a total between 150 and 200 ounces. An important byproduct output of sapphires, however, was made.

PRESENT AND PROPOSED PLACER MINING

CONFEDERATE CREEK

M. A. Ellis, who is now operating a dragline excavator on Indian Creek, west of Townsend, has acquired title to all of the valley of Confederate Creek, from its mouth upstream to Cement Gulch. He owns also 1320-foot claims at the lower ends of Cement, Montana, and Greenhorn Gulches. After completing the placer mining on Indian Creek (probably about 1948) Mr. Ellis plans to install a dragline plant on Confederate Creek; and starting about 1½ miles upstream from the "doodle-bug" tailings, he will work progressively upstream to Cement Gulch, and up the lower valleys of Greenhorn,

Montana, and Cement Gulches. If this venture proves profitable, Ellis and his partner, A. R. Douglas, will later install a bucket dredge in the lower valley of Confederate Creek. It is estimated by these operators that 30,000,000 cubic yards of workable gravel remain in this valley.

WHITE GULCH

A dragline plant was installed in August 1946, to work the stream gravels of White Gulch. Starting about 0.9 mile upstream from Number 16 Gulch, a 700-foot bedrock drain was dug, to get to bedrock. At the upper end of this drain, where mining begins, the depth to bedrock is estimated to be 35 to 40 feet.

The excavator is a No. 411 "Walking Page" monighan, made by the Page Engineering Co. The power consists of a diesel engine rated at 120 horsepower, which uses 35 gallons of fuel oil in 24 hours. This machine has two spuds in front, and one behind, each of which has shoes, the rear shoe being the largest. The spuds elevate the excavator, and after its elevation it is moved forward or backward, as needed. A boom, 60 feet in length, swings a bucket with a capacity of 1¾ cubic yards. Three winches are used, two for the boom and shovel, and one for moving the excavator. The washing plant and pump had not been delivered on the ground by the fall of 1946. Fred Gooden of Los Angeles. Calif., is the owner and operator of this plant.

The valley floor of White Gulch, southwest of the Big Belt Mountains, is not known to have been prospected, but it is possible that a low-grade pay streak exists on this foreland. If this were ever mined, the work would be done probably by dredging. About 3 miles of this stream course will lie above the level of the lake that is to be formed by the new Canyon Ferry Dam; and as dredging can be done under water, the dam will cause no difficulty. In fact, as this part of the stream bed is dry, the closer proximity of water would aid the operation, as pumping would have to be used for dredge flotation.

HELLGATE GULCH

During the summer of 1946, two men attempted to work a small open-cut on Hellgate Gulch, about 6 miles from its mouth, and about 1,000 feet upstream from the north side of the Flathead quartzite. The operation was not successful, but the work uncovered some old mining timbers, showing that drift mining or prospecting had once been done at this site.

CAVE GULCH AND COOPER GULCH

The foreland southwest of Cave Hill and Magpie Bluff is now agricultural land, which has not been available for mining. After the new Canyon Ferry dam is built, this land will be flooded. If a pay streak continues from Cave Gulch and Cooper Gulch onto this foreland, it is possible that this ground could be dredged. The feasibility of such a project would have to be determined by drilling, and by the rate of flooding on this land after the completion of the new dam.

OREGON GULCH AND CLARKS GULCH

One man worked on a very small scale for about 6 weeks in 1946, trying to mine the bench gravels along the southeast side of Oregon Gulch, near the mouth of Clarks Gulch, but this work was unsuccessful. Two other men were preparing to work the terrace gravels in the southwest side of Clarks Gulch, about a quarter of a mile above its junction with Oregon Gulch.

Above the canyon on Oregon Gulch, William H. De Borde has been working on a small scale for 3 or 4 years . About 2,000 feet upstream from the mouth of Holiday Gulch, a high terrace along the west side of Oregon Gulch was extensively worked in the early days of mining, as was also the bed of the creek. Between these two workings is a narrow intermediate bench, which was either overlooked, or was incompletely mined. Here, about 50 yards from the creek, Mr. De Borde has sunk several shafts, and has driven drifts from them. At this site the section consists of 2 to 4 feet of auriferous gravel, overlain by 6 feet of hillside wash, in turn overlain by 6 feet of tailings from the older terrace workings. The gravels are fine and fairly well rounded, and the gold occurs mainly on the surface of bedrock, though some gold is found in the lower foot or two of gravel. In places, however, it is necessary to mine all the gravel to obtain a high recovery of gold. The fineness of the gold is about 830.

MISSOURI RIVER TERRACES

The dredge operated by the Perry-Schroeder Mining Co. on Eldorado Bar is now idle on the upper terrace, where it last worked. The company is looking for a new site for dredging in this vicinity, and for this reason were prospecting, during the summer of 1946, along the bars of the Missouri, downstream from Canyon Ferry, and on the submerged terrace ground offshore from French Bar. The results of this prospecting are not known to the writer. It is possible that dredging may be attempted along this stretch of Hauser Lake.

ORIGIN OF THE PLACERS

SOURCES OF THE GOLD

The principal gold lodes in the York district lie in the valley of Trout Creek, above the mouth of York Gulch, and in York Gulch near the mouth of Rattlesnake Gulch. No pay streak, however, has been found in Trout Creek above the mouth of York Gulch; and the placers of York Gulch continue upstream above the principal lodes, to the heads of Rattlesnake Gulch and Kingsbury Gulch. This distribution of placers indicates a major source of gold on the flanks of Hedges Mountain; and the placers of Oregon Gulch and Cave Gulch confirm this interpretation. No workable lodes, however, have been found on Hedges Mountain.

The lack of a pay streak on upper Trout Creek may be explained by the assumption that the gold lodes in that area have not been bared to erosion long enough to provide the necessary gold. But the situation in the valley of York Gulch requires a different explanation. One or two conditions must have prevailed. The gold lodes on Hedges Mountain may be widespread but so low grade that they cannot be developed as mines; or the lodes, either high or low grade, that originally existed may have been partly or completely eroded. Certain data that might have a bearing on this question, such as the fineness of the placer gold, and of the gold in the developed lodes, can no longer be obtained. But the high elevation placer at the extreme head of Hellgate Gulch shows that some gold lodes have been bared to erosion since early Tertiary times. Therefore, though no real proof can be offered, it is believed that the placers of York Gulch have been in the process of concentration and reconcentration for a very long time; and the explanation is favored that all or most of the original lode material on Hedges Mountain has been removed by erosion. This explanation will account also for the absence of workable lodes at the heads of Oregon Gulch, Cave Gulch, and Avalanche Creek. The scarcity of placer gold in the valley of Hellgate Gulch must result from the small degree of gold mineralization at the head of that stream.

The pay streak on Johnny Gulch is the head of the pay streak on White Gulch; and the pay streaks on Montana Gulch and Cement Gulch are the heads of the pay streak on Confederate Creek. These two gulches head in Miller Mountain. Therefore no conflict exists regarding the sources of the gold, and the beginnings of the several pay streaks. As in the York area, the placer gold and the gold now being recovered from the lodes cannot be compared; but the enormous volume of gold in the valley of Confederate Creek points to long continued erosion, concentration and reconcentration. The high-terrace gold on Boulder Bar also also suggests the same origin. No conclusions, however, can be drawn regarding the character and tenor of the lodes eroded from Miller Mountain, as compared with the uneroded lodes that still remain.

OUTLINE OF PHYSIOGRAPHY

The formation of pay streaks is inextricably coupled with the physiographic history of the Missouri River and the Big Belt Mountains. The earlier chapters of this history are obscure and incomplete, but they can be neglected, as they constitute an unessential part of the story of the placers.

The earliest part of the significant physiographic history starts with an ancient erosional surface of Miocene age. This surface, or its reflection after long-continued residual and fluvial modification, is now a rolling terrain of moderate to low relief, which forms the top of the Big Belt Mountains. The Pliocene history of the Big Belt Mountains is obscure, because no erosional relics, or terraces, of that age have been recognized. The best example of an early Pleistocene erosional surface is the high gravel-covered terrace east of Winston, across which Beaver Creek has carved a course to the Missouri. Cogent evidence indicates that the deposition of the gravels on this terrace was concurrent with ancient glaciation in the Elkhorn and Big Belt Mountains. Subsequently, in mid-Pleistocene time, the streams draining the Big Belt Mountains were adjusted to a base level that was about 150 to 200 feet above the present level of the Missouri River. During this period of stability a pediment was developed southwest of the Big Belt Mountains, on the upturned and eroded edges of the Tertiary lake beds. Workable placers, if they were formed on this frontal pediment, have not been preserved.

At some later stage of the Pleistocene, starting in mid-Pleistocene time and extending probably to late Pleistocene time, the local base level of erosion was rapidly lowered to an altitude about 50 feet above the present level of the Missouri River. This sudden change in base level initiated rejuvenation in the streams draining the southwest slopes of the Big Belt Mountains; and this rejuvenation has worked upstream for various distances, depending upon the water supply and other factors. The adjustment is nearly complete in the valleys of Trout Creek and Confederate Creek, but in the valleys of less vigorous streams, such as Oregon, White, Cave, and Hellgate Gulches, and Magpie and Avalanche Creeks, the adjustment is incomplete, with the result that canyons or constricted stretches with high gradients are present upstream from their outlets onto the pediment that flanks the Big Belt Mountains. Upstream from these stretches the streams flow in hanging valleys with gradients that are still adjusted to an ancient base level of erosion.

This general rejuvenation incised the lower valleys of the streams draining the Big Belt Mountains, and caused a progressive reworking of preformed bench and stream placers. Gold thus acquired was deposited in the lower valleys of these streams, forming a series of placers ranging in age from the earliest one, formed in shallow valleys atop the pediment, to others occupying sites at various altitudes along the valley walls of the present streams.

In Recent time, another rapid lowering of the local base level of erosion took place, with the result that the Missouri River now flows at an altitude about 50 feet lower than its valley floor during late Pleistocene time. This likewise produced a rejuvenation which for the larger streams was largely completed in the foreland between the river and the face of the Big Belt Mountains. Incomplete adjustment of stream gradients is noticeable only in certain small gulches that head in the Tertiary foreland. Dry Hollow, with a sharp gorge at its mouth, is one of the best examples.

FORMATION OF PAY STREAKS

Pay streaks of several ages are distinguishable in the Big Belt Mountains. The oldest is represented by the high gravels in the extreme headwaters of Hellgate Gulch, at an altitude close to 7,000 feet, or about 3,300 feet above the level of the Missouri River. This pay streak consists of old stream gravels which were laid down in some ancient drainage channel upon an erosional surface of Miocene age. After the deposition of these gravels, late Tertiary lavas were discharged, with the result that some of the gravels were covered by lava and shielded from erosion. The high gravels at the head of Hellgate Gulch, originally preserved in this way, have now been bared to erosion. They are gold bearing, and therefore are classified as an ancient pay streak.

The general courses of the upper stream valleys draining the Big Belt Mountains were determined in early Pleistocene time. Most of the upper valleys are hanging valleys, separated by stretches of steeper gradient from their lower valleys; but the zones of separation are not of the same length or gradient. Other valleys, such as those of Trout Creek and Confederate Creek, are largely adjusted to the Pleistocene rejuvenation, and have no clearly recognizable zones of steepened gradient. Correlation of the placers in valleys adjusted, and in those unadjusted, to the rejuvenation is difficult, particularly without the aid of a topographic map. The correlations suggested are therefore tentative, and subject to revision.

The oldest Pleistocene placers are thought to be the bench placers that occur in unadjusted streams, above their zones of steepened gradient. Examples are the bench placers in the upper valleys of Oregon Gulch, and Avalanche Creek, above the mouth of Thompson Gulch. Correlative with them are the placers of Whites Bar, on White Gulch.

The several bench placers at different altitudes on Confederate Creek might be assumed superficially to correlate with much younger placers in lower valleys, such as those in the lower valley of Magpie Creek. But the terrace deposits of Confederate Creek are considered to be related to an old valley floor which is so far upstream that it has been little lowered by rejuvenation. It is thus conceived that the adjustment of Confederate Creek to the Pleistocene and Recent base levels of erosion has taken place largely downstream from Montana, Diamond, and Boulder Bars, and that the placers on those bars therefore correlate with bench placers in the upper valleys of Oregon Gulch and Avalanche Creek. An alternative interpretation might be that the placers of Montana and Diamond Bars are correlative with the undisturbed stream placers of Oregon Gulch and Avalanche Creek, above their zones of oversteepened gradient, and that Boulder Bar is to be correlated with the bench placers in those upper valleys. The preparation of a good topographic map of this district should throw much light on this problem.

The bedrock floors of the hanging valleys have not been greatly modified since the time of their formation, but the gravels overlying these bedrock surfaces have to various degrees been reworked, so that they range in age from the mid-Pleistocene to Recent. The lower levels of these gravels, however, have probably been little disturbed; and as the gold occurs close to bedrock, they constitute a part of the second or next younger group of Pleistocene placers. Examples are the placers on the floors of the upper valleys of Oregon, Cave, and Rattlesnake Gulches, and on Avalanche Creek. The placers in the valley floor of White Gulch, downstream from Whites Bar, probably constitute another example.

Downstream from the zones of steepened gradients, the streams draining the Big Belt Mountains differ in regard to the pay streaks that they contain. Hellgate and White Gulches and Avalanche Creek have neither stream nor bench placers in this part of their courses; but Oregon and Cave Gulches have bench placers at several levels. Magpie Creek also has placers at several levels, and in addition a pay streak in the stream gravels near its mouth. York Gulch has both stream and bench placers.

The oldest of the bench placers in the lower reaches of these streams were formed in shallow valleys on the pediment, at the time when rejuvenation of these valleys was just beginning. Examples are the placers on Cave Hill, atop the ridge west of Cave Gulch, and the placers of Centennial Bar, on lower Oregon Gulch. These deposits, though somewhat younger than the undisturbed stream placers of the upper valleys, are correlated roughly with them. This group of stream placers in the upper valleys, and uppermost bench placers in the lower valleys, therefore constitute the second, or intermediate group of Pleistocene placers.

Below the highest bench placers in the lower valleys of the streams, at the west end of the quadrangle, are the bench placers at lower altitudes. Examples are the placers at several levels in the lower valleys of Cave, Cooper, and Oregon Gulches, and in the lower valley of Magpie Creek. They were formed at different times throughout the period when these streams were undergoing readjustment to a base level 150 feet lower than the surface of the pediment flanking the Big Belt Mountains. Possibly some remnants of low terraces in the valley of Confederate Creek should be correlated with these. This group of bench deposits, at different altitudes and of different ages, are classed together as the youngest of the Pleistocene pay streaks.

After the last lowering of the base level of erosion, the streams continued to rework and redeposit the older gravels, producing on some streams pay streaks in the present valley floors. The pay streak at the mouth of Magpie Creek, that was mined by dredging, is an example. Confederate Creek and York Gulch also have pay streaks of Recent age. All these auriferous gravels in the present valley floors, however, which lie downstream from hanging valleys, or on streams where no hanging valleys exist, are grouped together as the fifth, or youngest of all the pay streaks. They are essentially post-Pleistocene in age.

NONMETALLIFEROUS DEPOSITS

The nonmetalliferous deposits of the Canyon Ferry quadrangle, were studied by an appraisal party of the Geological Survey, consisting of Robert M. Dreyer and Alfred L. Bush. The principal resources listed in this report are dimension stone, road metal, concrete aggregate, riprap, lime, cement rock, dolomite, bentonite, and sapphires. A part of the data on these deposits, exclusive of the sapphires, is taken from the work of Dreyer and Bush.

DIMENSION STONE

Two types of igneous rocks are available in the Canyon Ferry quadrangle that might be usable as dimension stone. One, a coarse-grained granitic rock, crops out along the road close to Canyon Ferry, and is well-exposed at an old quarry near the Canyon Ferry dam. This rock was used in the construction of the dam. It consists of a granular granitic groundmass set with large tabular phenocrysts of potash feldspar. Biotite is the principal dark mineral, but other dark minerals are present. The over-all color is gray, and the rock should yield a good grade of unfinished or finished building stone. It is probably too much fractured to be quarried as a monumental stone.

Farther south in the Spokane Hills, and also in the southwestern part of the quadrangle, this intrusive is a granular nonporphyritic quartz monzonite, containing about the same minerals as the granite, but in different proportions. The quartz monzonite should be useful for unfinished or finished building stone, curbing and similar applications. It is possible also that it would yield desirable monumental stone, but this application would have to be determined by a study of its jointing, sheeting, and natural planes of fracture.

Several limestones are present in the Big Belt Mountains and the Spokane Hills, of which the Madison limestone is probably the most desirable as a building stone. This formation consists of two members. In the Big Belt Mountains, a part of the upper member is massive and imperfectly bedded, whereas another part is evenly bedded, but weathers as massive rock. The lower member is thin-bedded, and weathers to slabs. The limestone of both members is dominantly light gray. The limestone of the lower member could be used where little trimming is required. The evenly bedded, but massive, limestone of the upper member might be used where larger blocks would be needed, and more splitting and cutting could be afforded. Both of these limestones crop out near the southwestern face of the Big Belt Mountains, and are therefore readily accessible.

The Madison limestone of the Spokane Hills has undergone considerable thermal metamorphism, so that the upper member is everywhere a coarse-grained, friable, white marble, that probably would be undesirable as a building stone. Much of the limestone of the lower member likewise is marmorized, but some blue-gray beds occur that might be useful. None of these limestones or marbles, however, is near a road and for that reason they are unavailable.

BROKEN STONE

Crushed and broken stone is used for many purposes, of which a few are road metal, concrete aggregate, railroad ballast, and riprap. Any of the limestones, quartzites, or igneous rocks of the quadrangle should make satisfactory road metal, but the upper Madison limestone, the Jefferson limestone, and the igneous rocks are massive rocks, which would have to be quarried and crushed. The lower Madison, Pilgrim, and Meagher limestones, however, could be more easily quarried, and are accessible along the face of the Big Belt Mountains. Gravels suitable for road metal are available at many places, but most of them would have to be screened to remove the larger boulders. The most readily accessible are the bench gravels exposed by placer mining in the cliffs at the end of the spur between Cave Gulch and Cooper Gulch. Stream gravels are available in the bed of Magpie Creek, where they have been bared by dredging. Large quantities of gravel, likewise bared by mining operations, are accessible on French and Eldorado Bars. Similarly at the south end of the quadrangle a large quantity of washed gravel can be obtained from the dredging dumps of Confederate Creek.

The same rocks that could be used for road metal would be satisfactory also for concrete aggregate, except that limestones would have to be selected that are free from beds and laminae of shale. Double sieving of the gravels at the mouth of Cave Gulch would be required, because they contain many muddy beds. The stream gravels on Magpie Creek, and the bench gravels of French and Eldorado Bars have been washed free of fine sediment, and would require only a single sieving for classification to a suitable size. The same is true of the gravels bared by dredging on Confederate Creek. Most of the larger boulders are quartzite. These would be suitable for riprap.

So far as known, only the Madison limestone is sufficiently pure to be used in making quicklime and hydrated lime. This limestone is being mined for these purposes at Elliston, about 20 miles west of Helena. The suitability of the Madison or other limestones for the manufacture of Portland cement would have to be determined by chemical analysis.

The Jefferson limestone is dolomitic, as is also the upper part of the Pilgrim limestone. These formations are readily accessible at the front of the Big Belt Mountains. The Helena and Newland limestones, of the Belt series, also are dolomitic, but they occur farther northeast in the Big Belt Mountains, and are relatively inaccessible. No analyses of these several dolomites, to determine their suitability for various chemical purposes, have yet been made.

BENTONITE

Bentonite occurs in unit 2 of the Tertiary sequence. The principal locality is north and south of Beaver Creek, from 3 to 5 miles northeast of Winston. The deposits were studied by Dreyer and Bush, and the field data here given are largely the result of their work.

The beds of unit 2 occur northwest of Beaver Creek and east of the Spokane Hills; they extend southeast of Beaver Creek for a short distance, beyond which they are covered by terrace gravels. The beds of bentonite occur near the base of this Tertiary unit, and the bentonization of the ash beds diminishes both eastward, across the strike, and northward from Beaver Creek. South of Beaver Creek the bentonized beds disappear below terrace gravels. The total distance along the strike, wherein the bentonite occurs, is about 3 miles; and the distance across the strike varies from a half to three-quarters of a mile. The best bentonite occurs within a mile north and south of Beaver

Creek. A sketch map prepared by Dreyer and Bush, is presented herewith as plate 2.

The beds of bentonite are commonly brownish green, but locally are buff, tan, and red. Repeated wetting and drying at the surface has resulted in expansion and contraction, that has produced the characteristically cracked appearance. The factor of swelling is about two. The tendency of this material to flow when it is wet has caused it to spread laterally over adjacent beds, so that the stratigraphic limits of the bentonite are not well defined. This bentonite consists partly of montmorillonite, and partly of other material that is not subject to bentonization. Therefore all of the beds are impure. Also they grade both along and across the strike into other beds that show less, or even no appreciable alteration. For all these reasons, the zones of bentonite that are mapped in plate 2 are less definite than their mapping indicates. They represent the best judgment of Dreyer and Bush regarding the sites where commercial deposits might be available.

Only the principal beds or zones of bentonite are shown in plate 2. Others that are too small to map have not been shown, and still others that are considered too impure to be useful have been omitted. The width of each of these zones is particularly indefinite because of surficial flowage, so that it is probable that the beds will be thinner below the surface than they appear on the map. It is nevertheless true that a large volume of low-grade bentonite is present, that would be useful as a sealing agent for dams and canals, and other purposes for which a low-swelling material is satisfactory. Gravel roads along both sides of Beaver Creek render these deposits easily accessible.

SAPPHIRES

DISCOVERY AND EARLY MINING

The earliest reference to sapphires in Montana, according to Kunz (1899, p. 692), was the discovery of such gem stones by E. R. Collins, on May 5, 1865, at some unspecified locality; but Browne (1868, p. 502) mentions specifically that they were known in the alluvial deposits of Eldorado Bar by 1867. In 1889, according to Kunz (1893, pp. 542–544), a British company, called the Montana Sapphire and Ruby Co., Ltd., was organized to mine the sapphires on Eldorado Bar and other bars of the Missouri River farther upstream. This company was capitalized at 450,000 shares of stock, which were offered at one pound a share; and it purchased 4,000 acres of Eldorado Bar from Augustus A. Spratt, Frank D. Spratt, and other owners of this ground. One subsidiary called the Spokane Sapphire Co., and also independent companies, were organized. For a few years, both before and after the formation of the Montana Sapphire and Ruby Co., Ltd., a small output of sapphires of gem-stone grade was made, but the project did not flourish, and in 1893 the company failed. Sapphires were mined also by one of these early companies on a terrace that lies south-southwest from Canyon Ferry.

Kunz (1890, pp. 48, 49) states that the value of the gems cut from material found in this district amounted at one time to $2,000 a year. He mentioned also among the recovered sapphires an oriental emerald of 3½ carats, other gems weighing from 4 to 6 carats, and one steel-blue stone of 9 carats. He further states that a crescent-shaped piece of jewelry was made from these stones by Tiffany & Co. in 1883.

DISTRIBUTION

Sapphires have been found at many places along the Missouri River, from the vicinity of Canyon Ferry downstream to and probably beyond Eldorado Bar. Most of these occurrences are in unconsolidated deposits, but a few are in bedrock.

The southeasternmost known limit of the sapphires is at the mouth of Magpie Creek, where small amounts are reported to have been found in the concentrates recovered by dredging. Sapphires were found also in the low-grade auriferous gravels at the southwest end of the spur between Cave Gulch and Cooper Gulch, and in the terrace and stream gravels of Oregon Gulch, close to the Missouri River. The largest volume of sapphires, however, was found on the bars and terraces of the river, from French Bar to Eldorado Bar.

A gravel deposit containing sapphires is situated on a high terrace south-southwest of Canyon Ferry, about 240 feet above the level of the river. The bedrock at this place is decomposed granitic rock, which is overlain by ancient river gravels ranging in thickness from a few feet to 10 feet. These gravels are well rounded, and have a maximum size of one foot in diameter. They consist mainly of andesitic volcanics, Flathead quartzite, rocks of the Belt series, and some of the local granitic bedrock. Sapphires were evidently found in these gravels, which were explored by a number of shallow shafts and inclined drifts. A tunnel also was driven into the gravels at the face of this terrace. All these old workings are now caved. It is reported that the gravels from this site were hauled to the river, where they were washed for their content of sapphires.

Sapphires have been found also in place at several localities. Kunz (1890, p. 49) mentions a specimen, shown to him, of a trachytic dike in which were embedded well-defined crystals of sapphire. The exact locality was not stated, but it was said to be along the Missouri River near and above Eldorado Bar. A more specific reference was made by Kunz (1893, pp. 543–544) to 2 dikes found on Ruby Bar, which was said to be 6 miles upstream from Eldorado Bar, (evidently between French Bar and Gruel Bar). A third bedrock locality is at French Bar, where according to Benjamin T. Smith,[4] of York, the Montana Sapphire and Ruby Co., Ltd., drove several tunnels wherein one or more sapphire-bearing dikes were discovered. These tunnels are now submerged by the water of Hauser Dam, but in one of these dikes, visible at the surface, crystals of sapphire occur sparingly.

CHARACTER AND ORIGIN

The sapphires recovered in recent years by gold-dredging operations on Eldorado Bar are predominantly tabular six-sided crystals with rounded edges. Most of them are light green or light blue, but some are nearly colorless, yellow, lavender, or pink. The concentrates recovered by the dredge were classified from 2-mesh to 8-mesh, and the sapphires were separated from these fractions. Few, if any, crystals larger than three quarters of an inch in size were found, and no attempt was made to save the material smaller than 8-mesh. Relatively few stones of gem quality were recovered.

The first dike found on Ruby Bar contained visible crystals of sapphire, garnet, and sanadine. The second dike, bared by mining operations, was a vesicular mica-augite andesite containing phenocrysts of brown mica and augite, in a matrix of feldspar microlites, interstitial glass and much magnetite. The sapphire-bearing dike on French Bar is a light-gray finely porphyritic rock, consisting of phenocrysts of plagioclase and biotite, in a matrix of minute needles of plagioclase and glass. The phenocrysts of plagioclase range in size to a maximum of 3 millimeters; the phenocrysts of biotite are smaller. The plagioclase phenocrysts are almost altered to carbonates, but some of the plagioclase in the matrix is unaltered, and appears to be andesine. Sapphires are rare, but where present are commonly as large or larger than the phenocrysts of feldspar. Some crystals of garnet also are present. The rock is probably an andesite porphyry.

Some of the earlier writers have assumed that dikes of this kind are the principal sources of the sapphires along this stretch of the Missouri River. The volume of sapphires so far recovered, however, together with the discarded and unrecovered material, is known to aggregate many tens of thousands of ounces. Moreover, these dikes, with or without sapphires, are rare; and as they originate at depth, they cannot be assumed to have been more plentiful at the Pleistocene or Pliocene surface than they appear to be at the present surface. Such dikes are therefore regarded as totally inadequate sources of the sapphires.

---

[4] Smith, Benjamin T., oral communication.

The origin is indicated by the horizontal and vertical distribution of the sapphires, and by the local geology near Canyon Ferry. The sapphires, as heretofore noted, are found in the stream gravels of Magpie Creek, and in the high-terrace gravels on the southwest side of the Missouri River, opposite Magpie Creek. They are not reported anywhere upstream from this point on the Missouri, and are found everywhere downstream, both in stream and terrace gravels. Evidently therefore the bedrock source did not lie southeast of Magpie Creek; and judging by the occurrence of sapphires in the highest river terraces, the bedrock source was bared to erosion before the beginning of the Pleistocene epoch. The country rock along the southwest side of the Missouri River, at and near Canyon Ferry, is pre-Cambrian shale and Cambrian limestone, that are intruded by a monzonitic pluton, with an elongation trending northwestward. At its northwestern end, near Canyon Ferry, this intrusive rock is characterized by numerous variations in fabric and composition, such that several variants are present that differ materially from the normal type of quartz monzonite farther to the southeast. It is believed that these variants of the quartz monzonite, in contact with the country rock, and particularly in contact with limestone, produced contact-metamorphic deposits, wherein a large volume of corundum was formed. Such deposits apparently were formed along the roof of the intrusive; and in the long interval since they were bared to erosion have been entirely eroded. The total or part obliteration of bedrock sources is more common than otherwise in the apparent genesis of placers.

RECENT MINING

The dredging of auriferous gravels on Eldorado Bar between 1938 and 1944 has already been described. A byproduct of this mining was the production in 6½ years of about 58,000 Troy ounces of sapphires, suitable for jewel bearings. An equal or greater amount of lower-grade material also was produced, and sold for other uses. As no sapphires smaller than 8-mesh size were saved, the percentage of corundum per cubic yard, or per unit weight of concentrates, is not known. At an average price of $1.00 a Troy ounce, however, the tenor of the gravels in plus 8-mesh sapphires is estimated to have been about four-sevenths cent to the cubic yard, and the value of the output was probably between $60,000 and $75,000.

LITERATURE CITED

Browne, J. Ross, 1868, U. S. Treas. Dept., Mineral resources of the states and territories west of the Rocky Mountains, 1867.

Douglass, Earl, 1908, Fossil horses from North Dakota and Montana: Carnegie Mus. Annals, v. 4.

Kunz, George F., 1890, Gems and precious stones of North America: The Scientific Publishing Co., New York.

——— 1893, Precious stones: U. S. Geol. Survey, Mineral resources U. S., 1891.

——— 1894, Precious stones: U. S. Geol. Survey, Mineral resources U. S. 1893.

Lyden, Charles J., 1948, The gold placers of Montana: Montana Bur. Mines and Geol., Memoir 26.

Marshall, R. B., 1914, Profile surveys of Missouri River from Great Falls to Three Forks, Mont.: U. S. Geol. Survey Water-Supply Paper 367, 8 pp.

Pardee, J. T., 1925, Geology and ground-water resources of Townsend Valley, Mont.: U. S. Geol. Survey Water-Supply Paper 539, 61 pp.

Pardee, J. T. and Schrader, F. C., 1933, Metalliferous deposits of the greater Helena mining region, Montana: U. S. Geol. Survey Bull. 842, 318 pp.

U. S. Weather Bureau, 1936, Climatic summary of the United States, from the establishment of stations up to 1930: U. S. Dept. Agr.

INDEX

O

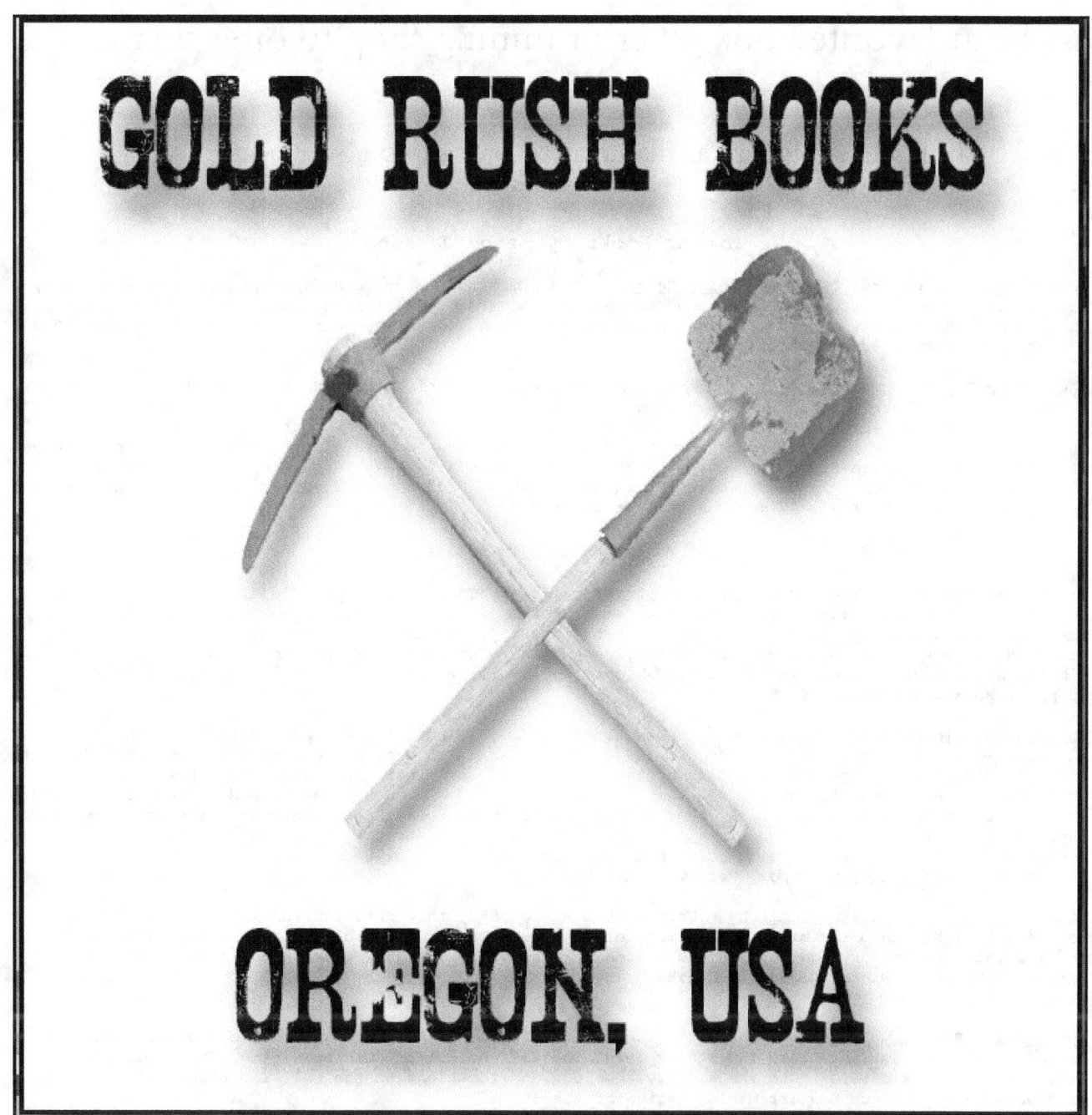

GOLD RUSH BOOKS

OREGON, USA

www.GoldMiningBooks.com

Books On Mining

Visit: www.goldminingbooks.com to order your copies or ask your favorite book seller or mining shop to offer them.

Note: Wholesale copies are available at 50% of listed cover price.

Mining Books by Kerby Jackson

<u>Gold Dust: Stories From Oregon's Mining Years</u> - Oregon mining historian and prospector, Kerby Jackson, brings you a treasure trove of seventeen stories on Southern Oregon's rich history of gold prospecting, the prospectors and their discoveries, and the breathtaking areas they settled in and made homes. **5" X 8", 98 ppgs. Retail Price: $11.99**

<u>The Golden Trail: More Stories From Oregon's Mining Years</u> - In his follow-up to "Gold Dust: Stories of Oregon's Mining Years", this time around, Jackson brings us twelve tales from Oregon's Gold Rush, including the story about the first gold strike on Canyon Creek in Grant County, about the old timers who found gold by the pail full at the Victor Mine near Galice, how Iradel Bray discovered a rich ledge of gold on the Coquille River during the height of the Rogue River War, a tale of two elderly miners on the hunt for a lost mine in the Cascade Mountains, details about the discovery of the famous Armstrong Nugget and others. **5" X 8", 70 ppgs. Retail Price: $10.99**

Alaska Mining Books

<u>Ore Deposits of the Willow Creek Mining District, Alaska</u> - Unavailable since 1954, this hard to find publication includes valuable insights into the Willow Creek Mining District near Hatcher Pass in Alaska. The publication includes insights into the history, geology and locations of the well known mines in the area, including the Gold Cord, Independence, Fern, Mabel, Lonesome, Snowbird, Schroff-O'Neil, High Grade, Marion Twin, Thorpe, Webfoot, Kelly-Willow, Lane, Holland and others. **8.5" X 11", 96 ppgs. Retail Price: $9.99**

<u>The Juneau Gold Belt of Alaska</u> - Unavailable since 1906, this hard to find publication includes valuable insights into the gold mines around Juneau, Alaska. The publication includes important details into the history, geology and locations of the well known gold mines and prospects in the area, including those around Windham Bay, Holkham Bay, Port Snettisham, on Grindstone and Rhine Creeks, Gold Creek, Douglas Island, Salmon Creek, Lemon Creek, Nugget Creek, from the Mendenhall River to Berners Bay, McGinnis Creek, Montana Creek, Peterson Creek, Windfall Creek, the Eagle River, Yankee Basin, Yankee Curve, Kowee Creek and elsewhere. Not only are gold placer mines included, but also hardrock gold mines. **8.5" X 11", 224 ppgs. Retail Price: $19.99**

<u>Mining in the Jumbo Basin of Alaska</u> - Unavailable since 1953, this hard to find publication includes valuable insights into the mines and geology of the Jumbo Basin. The publication includes important details into the history, geology and locations of the well known gold mines and prospects in the famous Jumbo Basin Mining Region of Alaska.
72 ppgs, 9.99

<u>The Rampart Placer Gold Region of Alaska</u> - Unavailable since 1906, this hard to find publication includes valuable insights into the placer gold mines of the Rampart Mining Region. The publication includes important details into the history, geology and locations of the well known gold mines and prospects in the famous Rampart Mining Region of Alaska.
78 ppgs, 10.99

Arizona Mining Books

Mines and Mining in Northern Yuma County Arizona - Originally published in 1911, this important publication on Arizona Mining has not been available for over a hundred years. Included are rare insights into the gold, silver, copper and quicksilver mines of Yuma County, Arizona together with hard to find maps and photographs. Some of the mines and mining districts featured include the Planet Copper Mine, Mineral Hill, the Clara Consolidated Mine, Viati Mine, Copper Basin prospect, Bowman Mine, Quartz King, Billy Mack, Carnation, the Wardwell and Osbourne, Valensuella Copper, the Mariquita, Colonial Mine, the French American, the New York-Plomosa, Guadalupe, Lead Camp, Mudersbach Copper Camp, Yellow Bird, the Arizona Northern (Salome Strike), Bonanza (Harqua Hala), Golden Eagle, Hercules, Socorro and others. **8.5" X 11", 144 ppgs. Retail Price: $11.99**

The Aravaipa and Stanley Mining Districts of Graham County Arizona - Originally published in 1925, this important publication on Arizona Mining has not been available for nearly ninety years. Included are rare insights into the gold and silver mines of these two important mining districts, together with hard to find maps. **8.5" X 11", 140 ppgs. Retail Price: $11.99**

Gold in the Gold Basin and Lost Basin Mining Districts of Mohave County, Arizona - This volume contains rare insights into the geology and gold mineralization of the Gold Basin and Lost Basin Mining Districts of Mohave County, Arizona that will be of benefit to miners and prospectors. Also included is a significant body of information on the gold mines and prospects of this portion of Arizona. This volume is lavishly illustrated with rare photos and mining maps. **8.5" X 11", 188 ppgs. Retail Price: $19.99**

Mines of the Jerome and Bradshaw Mountains of Arizona - This important publication on Arizona Mining has not been available for ninety years. This volume contains rare insights into the geology and ore deposits of the Jerome and Bradshaw Mountains of Arizona that will be of benefit to miners and prospectors who work those areas. Included is a significant body of information on the mines and prospects of the Verde, Black Hills, Cherry Creek, Prescott, Walker, Groom Creek, Hassayampa, Bigbug, Turkey Creek, Agua Fria, Black Canyon, Peck, Tiger, Pine Grove, Bradshaw, Tintop, Humbug and Castle Creek Mining Districts. This volume is lavishly illustrated with rare photos and mining maps. **8.5" X 11", 218 ppgs. Retail Price: $19.99**

The Ajo Mining District of Pima County Arizona - This important publication on Arizona Mining has not been available for nearly seventy years. This volume contains rare insights into the geology and mineralization of the Ajo Mining District in Pima County, Arizona and in particular the famous New Cornelia Mine. **8.5" X 11", 126 ppgs. Retail Price: $11.99**

Mining in the Santa Rita and Patagonia Mountains of Arizona - Originally published in 1915, this important publication on Arizona Mining has not been available for nearly a century. Included are rare insights into hundreds of gold, silver, copper and other mines in this famous Arizona mining area. Details include the locations, geology, history, production and other facts of the mines of this region. **8.5" X 11", 394 ppgs. Retail Price: $24.99**

Mining in the Bisbee Quadrangle of Arizona - Originally published in 1906, this important publication on Arizona Mining has not been available for nearly a century. Included are rare insights into hundreds of gold, silver, copper and other mines in this famous Arizona mining area. Details include the locations, geology, history, production and other facts of the mines of this important mining region. **8.5" X 11", 188 ppgs. Retail Price: $14.99**

Placer Gold Mining in Arizona - Unavailable since 1922, this hard to find publication includes valuable insights into the placer gold mines of the Arizona. Originally released as "Placer Gold of Arizona", despite its small size, this publication includes important details into the history, geology and locations of the well known placer gold mines and prospects in the State of Arizona. **48 ppgs, 8.99**

Gold and Copper Mining near Payson, Arizona - Written in 1915, this hard to find publication includes valuable insights into the gold and copper mining industry of Arizona. Highlighted here are the gold and copper mines near Payson, Arizona. **68 ppgs, 8.99**

Lode Gold Mining in Arizona - Unavailable since 1934, this hard to find publication, originally released as "Arizona Lode Gold Mines and Gold Mining" includes valuable insights into the gold mining industry of Arizona. Included are valuable insights into over 150 hardrock gold mines in over 30 different mining districts in Arizona. **278 ppgs, 21.99**

Mining in the Dragoon Quadrangle of Cochise County, Arizona - Unavailable since 1964, this hard to find publication includes valuable insights into the mines of the Dragoon Quadrangle Mining Region. The publication includes important details into the history, geology and locations of the well known mines and prospects in this famous mining region of Arizona. **224 ppgs., 19.99**

Directory of Operating Mines in Arizona in 1915 - Unavailable since 1916, this hard to find publication includes valuable insights into the mines of Arizona. This small publication includes a complete list of the mines that were operating in the State of Arizona during 1915 and includes details such as general location, owners and some basic facts about each mining operation. **52 ppgs. 8.99**

Arizona Ore Deposits - Unavailable since 1938, this hard to find publication includes valuable insights into some ore deposits of Arizona. Included are valuable insights into the formation and characteristics of valuable ore deposits in the Jerome, Miami, Inspiration, Clifton, Morenci, Ray, Ajo, Eureka, Tombstone and Magma mining districts. Included are details into some of the major gold, silver and copper mines of these important Arizona mining areas. 160 ppgs, 14.99

Mining in Santa Cruz County, Arizona - Written in 1916, this hard to find publication includes valuable facts on the mines of this famous mining area, which is the oldest in Arizona. Included in this small booklet are hard to find details on the history and mines and prospects of this area. 54 ppgs, 7.99

The Mineral Industries of Arizona: A Brief History of the Development of Arizona's Mineral Resources - Written in 1962, this hard to find publication includes valuable facts about the Arizona mining industry. Included in this small booklet is a brief history of gold, silver and copper mining in Arizona. 54 ppgs, 7.99

Mining in Southern Yuma County, Arizona - Written in 1933, this hard to find publication includes valuable facts on the gold, silver and copper mines of this famous mining area. Included are the hard to find locations, histories and details of numerous mines in Yuma County, including the Silver Clip, Amelia, Revelation,Mendevil, Chloride, Cash Entry, Mandarin, Saxon, Princess, Hamburg, Silver King, Geronimo, Red Cloud, Black Rock, Pacific, Mandan, Silver Glance, Papago, Riverview, Hardt, Broadway, Jupiter, Annie, Flora Temple, Senora, Castle Dome, New Dil, Lady Edith, Big Dome, Yuma, Little Dome, Hull, Cleveland-Chicago, Adams, Mabel, Lincoln, Big Eye, Sheep, Keystone, King of Arizona, North Dstar, Geyser, Rand, IXL, Regal, C.O.D., Big Horn, Cemitosa, Alamo, Alnoah, Tunnel Springs, and dozens of others. 262 ppgs., 20.99

Geology of the San Manuel Copper Deposit of Arizona - Written in 1951, this hard to find publication includes valuable facts about this important copper mining area in Pinal County, Arizona. 98 ppgs, 9.99

Mining in the Sierrita Mountains of Pima County, Arizona - Written in 1921, this hard to find publication includes valuable facts on the mines of this famous mining area in Pima County. Included in this small booklet are hard to find details on the history and mines and prospects of this area. 54 ppgs, 8.99

Mining in the Cerbat Range, Black Mountains and Grand Wash Cliffs: Mohave County, Arizona - Written in 1909, this hard to find publication includes valuable facts on the Mines and Mineral Deposits in the Cerbat Range, Black Mountains and Grand Wash Cliffs of Mohave County, Arizona. Included are the hard to find locations and details on dozens of gold, silver and copper mines in this famous Arizona mining region. 254 ppgs, 24.99

Uranium Mining at the Dripping Spring Quartzite in Gila County, Arizona - Written in 1969, this hard to find publication includes valuable facts on the Mines and Mineral Deposits in the uranium mining area of Dripping Spring in Gila County, Arizona. Included are the hard to find locations, details and maps of uranium deposits in Gila County. 136 ppgs, 12.99

Arizona Gold Placers - Written in 1927, this hard to find publication includes valuable insights into the gold placer deposits of Arizona. Highlighted here are the details you need to find placer gold in Arizona, including the location of Arizona placer gold mines. 92 ppgs, 8.99

Geology of the Lower Gila Region of Arizona - Written in 1921, this hard to find publication includes valuable facts on the geology in Gila County, Arizona. Included in this small booklet are hard to find details on the geology of this area that will be of aid to prospectors, miners, rock hounds and geologic students. 46 ppgs, 7.99

The Wallapai Mining District of Mohave County, Arizona - Written in 1951, this hard to find publication includes valuable facts on the mines of this famous mining area in the Cerbat Mountains. Included in this small booklet are hard to find details on the history and mines and prospects of this area. 68 ppgs, 8.99

California Mining Books

The Tertiary Gravels of the Sierra Nevada of California - Mining historian Kerby Jackson introduces us to a classic mining work by Waldemar Lindgren in this important re-issue of The Tertiary Gravels of the Sierra Nevada of California. Unavailable since 1911, this publication includes details on the gold bearing ancient river channels of the famous Sierra Nevada region of California. 8.5" X 11", 282 ppgs. Retail Price: $19.99

The Mother Lode Mining Region of California - Unavailable since 1900, this publication includes details on the gold mines of California's famous Mother Lode gold mining area. Included are details on the geology, history and important gold mines of the region, as well as insights into historic mining methods, mine timbering, mining machinery, mining bell signals and other details on how these mines operated. Also included are insights into the gold mines of the California Mother Lode that were in operation during the first sixty years of California's mining history. 8.5" X 11", 176 ppgs. Retail Price: $14.99

Lode Gold of the Klamath Mountains of Northern California and South West Oregon - Unavailable since 1971, this publication was originally compiled by Preston E. Hotz and includes details on the lode mining districts of Oregon and California's Klamath Mountains. Included are details on the geology, history and important lode mines of the French Gulch, Deadwood, Whiskeytown, Shasta, Redding, Muletown, South Fork, Old Diggings, Dog Creek (Delta), Bully Choop (Indian Creek), Harrison Gulch, Hayfork, Minersville, Trinity Center, Canyon Creek, East Fork, New River, Denny, Liberty (Black Bear), Cecilville, Callahan, Yreka, Fort Jones and Happy Camp mining districts in California, as well as the Ashland, Rogue River, Applegate, Illinois River, Takilma, Greenback, Galice, Silver Peak, Myrtle Creek and Mule Creek districts of South Western Oregon. Also included are insights into the mineralization and other characteristics of this important mining region. 8.5" X 11", 100 ppgs. **Retail Price: $10.99**

Mines and Mineral Resources of Shasta County, Siskiyou County, Trinity County: California - Unavailable since 1915, this publication was originally compiled by the California State Mining Bureau and includes details on the gold mines of this area of Northern California. Also included are insights into the mineralization and other characteristics of this important mining region, as well as the location of historic gold mines. 8.5" X 11", 204 ppgs. **Retail Price: $19.99**

Geology of the Yreka Quadrangle, Siskiyou County, California - Unavailable since 1977, this publication was originally compiled by Preston E. Hotz and includes details on the geology of the Yreka Quadrangle of Siskiyou County, California. Also included are insights into the mineralization and other characteristics of this important mining region. 8.5" X 11", 78 ppgs. **Retail Price: $7.99**

Mines of San Diego and Imperial Counties, California - Originally published in 1914, this important publication on California Mining has not been available for a century. This publication includes important information on the early gold mines of San Diego and Imperial County, which were some of the first gold fields mined in California by early Spanish and Mexican miners before the 49ers came on the scene. Included are not only details on early mining methods in the area, production statistics and geological information, but also the location of the early gold mines that helped make California "The Golden State". Also included are details on the mining of other minerals such as silver, lead, zinc, manganese, tungsten, vanadium, asbestos, barite, borax, cement, clay, dolomite, fluospar, gem stones, graphite, marble, salines, petroleum, stronium, talc and others. 8.5" X 11", 116 ppgs. **Retail Price: $12.99**

Mines of Sierra County, California - Unavailable since 1920, this publication was originally compiled by the California State Mining Bureau and includes details on the gold mines of Sierra County, California. Also included are insights into the mineralization and other characteristics of this important mining region, as well as the location of historic gold mines. 8.5" X 11", 156 ppgs. **Retail Price: $19.99**

Mines of Plumas County, California - Unavailable since 1918, this publication was originally compiled by the California State Mining Bureau and includes details on the gold mines of Plumas County, California. Also included are insights into the mineralization and other characteristics of this important mining region, as well as the location of historic gold mines. 8.5" X 11", 200 ppgs. **Retail Price: $19.99**

Mines of El Dorado, Placer, Sacramento and Yuba Counties, California - Originally published in 1917, this important publication on California Mining has not been available for nearly a century. This publication includes important information on the early gold mines of El Dorado County, Placer County, Sacramento County and Yuba County, which were some of the first gold fields mined by the Forty-Niners during the California Gold Rush. Included are not only details on early mining methods in the area, production statistics and geological information, but also the location of the early gold mines that helped make California "The Golden State". Also included are insights into the early mining of chrome, copper and other minerals in this important mining area. 8.5" X 11", 204 ppgs. **Retail Price: $19.99**

Mines of Los Angeles, Orange and Riverside Counties, California - Originally published in 1917, this important publication on California Mining has not been available for nearly a century. This publication includes important information on the early gold mines of Los Angeles County, Orange County and Riverside County, which were some of the first gold fields mined in California by early Spanish and Mexican miners before the 49ers came on the scene. Included are not only details on early mining methods in the area, production statistics and geological information, but also the location of the early gold mines that helped make California "The Golden State". 8.5" X 11", 146 ppgs. **Retail Price: $12.99**

Mines of San Bernadino and Tulare Counties, California - From 1917, this publication on California Mining has not been available for nearly a century. This publication includes important information on the early gold mines of San Bernadino and Tulare County, which were some of the first gold fields mined in California by early Spanish and Mexican miners before the 49ers came on the scene. Included are not only details on early mining methods in the area, production statistics and geological information, but also the location of the early gold mines that helped make California "The Golden State". Also included are details on the mining of other minerals such as copper, iron, lead, zinc, manganese, tungsten, vanadium, asbestos, barite, borax, cement, clay, dolomite, fluospar, gem stones, graphite, marble, salines, petroleum, stronium, talc and others. 8.5" X 11", 200 ppgs. **Retail Price: $19.99**

Chromite Mining in The Klamath Mountains of California and Oregon - Unavailable since 1919, this publication was originally compiled by J.S. Diller of the United States Department of Geological Survey and includes details on the chromite mines of this area of Northern California and Southern Oregon. Also included are insights into the mineralization and other characteristics of this important mining region, as well as the location of historic mines. Also included are insights into chromite mining in Eastern Oregon and Montana. 8.5" X 11", 98 ppgs. Retail Price: $9.99

Mines and Mining in Amador, Calaveras and Tuolumne Counties, California - Unavailable since 1915, this publication was originally compiled by William Tucker and includes details on the mines and mineral resources of this important California mining area. Included are details on the geology, history and important gold mines of the region, as well as insights into other local mineral resources such as asbestos, clay, copper, talc, limestone and others. Also included are insights into the mineralization and other characteristics of this important portion of California's Mother Lode mining region. 8.5" X 11", 198 ppgs. Retail Price: $14.99

The Cerro Gordo Mining District of Inyo County California - Unavailable since 1963, this publication was originally compiled by the United States Department of Interior. Included are insights into the mineralization and other characteristics of this important mining region of Southern California. Topics include the mining of gold and silver in this important mining district in Inyo County, California, including details on the history, production and locations of the Cerro Gordo Mine, the Morning Star Mine, Estelle Tunnel, Charles Lease Tunnel, Ignacio, Hart, Crosscut Tunnel, Sunset, Upper Newtown, Newtown, Ella, Perseverance, Newsboy, Belmont and other silver and gold mines in the Cerro Gordo Mining District. This volume also includes important insights into the fossil record, geologic formations, faults and other aspects of economic geology in this California mining district. 8.5" X 11", 104 ppgs. Retail Price: $10.99

Mining in Butte, Lassen, Modoc, Sutter and Tehama Counties of California - Unavailable since 1917, this publication was originally compiled by the United States Department of Interior. Included are insights into the mineralization and other characteristics of this important mining region of California. Topics include the mining of asbestos, chromite, gold, diamonds and manganese in Butte County, the mining of gold and copper in the Hayden Hill and Diamond Mountain mining districts of Lassen County, the mining of coal, salt, copper and gold in the High Grade and Winters mining districts of Modoc County, gold mining in Sutter County and the mining of gold, chromite, manganese and copper in Tehama County. This volume also includes the production records and locations of numerous mines in this important mining region. 8.5" X 11", 114 ppgs. Retail Price: $11.99

Mines of Trinity County California - Originally published in 1965, this important publication on California Mining has not been available for nearly fifty years. This publication includes important information on mines and mining in Trinity County, California, as well insights into the mineralization and geology of this important mining area in Northern California. Included are extensive details on hardrock and placer gold mines and prospects, including charts showing the locations of these historic mines.. 8.5" X 11", 144 ppgs. Retail Price: $12.99

Mines of Kern County California - Originally published in 1962, this important publication on California Mining has not been available for nearly fifty years. This publication includes important information on mines and mining in Kern County, California, as well insights into the mineralization and geology of this important mining area in California. Included are extensive details on hardrock and placer gold mines and prospects, including charts showing the locations of these historic mines. 8.5" X 11", 398 ppgs. Retail Price: $24.99

Mines of Calaveras County California - Originally published in 1962, this important publication on California Mining has not been available for nearly fifty years. This publication includes important information on mines and mining in Calaveras County, California, as well insights into the mineralization and geology of this important mining area in Northern California. Included are extensive details on hardrock and placer gold mines and prospects, including charts showing the locations of these historic mines. 8.5" X 11", 236 ppgs. Retail Price: $19.99

Lode Gold Mining in Grass Valley California - Unavailable since 1940, this publication was originally compiled by the United States Department of Interior. Included are insights into the gold mineralization and other characteristics of this important mining region of Nevada County, California. This volume also includes important insights into the geologic formations, faults and other aspects of economic geology in this California mining district. Of particular interest are the fine details on many hardrock gold mines in the area, including their locations, histories, development and mineralization. Some of the mines featured include the Gold Hill Mine, Massachusetts Hill, Boundary, Peabody, Golden Center, North Star, Omaha, Lone Jack, Homeward Bound, Hartery, Wisconsin, Allison Ranch, Phoenix, Kate Hayes, W.Y.O.D., Empire, Rich Hill, Daisy Hill, Orleans, Sultana, Centennial, Conlin, Ben Franklin, Crown Point and many others. 8.5" X 11", 148 ppgs. Retail Price: $12.99

Lode Mining in the Alleghany District of Sierra County California - Unavailable since 1913, this publication was originally compiled by the United States Department of Interior. Included are insights into the mineralization and other characteristics of this important mining region of Sierra County. Included are details on the history, production and locations of numerous hardrock gold mines in this famous California area, including the Tightner Mine, Minnie D., Osceola, Eldorado, Twenty One, Sherman, Kenton, Oriental, Rainbow, Plumbago, Irelan, Gold Canyon, North Fork, Federal, Kate Hardy and others. This volume also includes important insights into the fossil record, geologic formations, faults and other aspects of economic geology in this California mining district. **8.5" X 11", 48 ppgs. Retail Price: $7.99**

Six Months In The Gold Mines During The California Gold Rush - Unavailable since 1850, this important work is a first hand account of one "49'ers" personal experience during the great California Gold Rush, shedding important light on one of the most exciting periods in the history of not only California, but also the world. Compiled from journals written between 1847 and 1849 by E. Gould Buffum, a native of New York, "Six Months In The Gold Mines During The California Gold Rush" offers a rare look into the day to day lives of the people who came to California to work in her gold mines when the state was still a great frontier. 8.5" X 11", 290 ppgs. Retail Price: $19.99

Quartz Mines of the Grass Valley Mining District of California - Unavailable since 1867, this important publication has not been available since those days. This rare publication offers a short dissertation on the early hardrock mines in this important mining district in the California Mother Lode region between the 1850's and 1860's. Also included are hard to find details on the mineralization and locations of these mines, as well as how they were operated in those day. 8.5" X 11", 44 ppgs. Retail Price: $8.99

Gold Rush on the Feather River - First published in 1924, this short publication by G.C. Mansfield sheds important light on the early history of gold mining on the Feather River. Included are rare insights into the first decade of gold mining and the early mining camps of the Feather River during the 1850's. 64 ppgs., 9.99

The Bodie Mining District of California - First published in 1986, it has been unavailable since those days and sheds important light on this famous mining area. Included are the history, characteristics and locations of numerous old mines around the ghost town of Bodie. 64 ppgs, 8.99

Geology and Mineral Resources of the Gasquet Quadrangle of California-Oregon - First published in 1953, it has been unavailable for over a century and sheds important light on the geological features and mineral resources of this portion of Northern California and Southern Oregon. 80 ppgs, 9.99

Gold Dredging in California - Unavailable since 1905, this publication was originally compiled by the California Bureau of Mines. A century and more ago, giant dredging machines dug in California's rivers and creeks in search of illusive golden riches. First appearing in the 1850's, gold dredges finally reached their peak of development in Siberia and New Zealand before becoming popular again in the United States. This book offers a unique historical perspective on the gold dredges that once operated in California. This book on California mining history is lavishly illustrated with dozens of rare historic photos gold dredges that once operated in California, as well as hard to locate plans on how these dredges were designed. 148 ppgs, 12.99

Gold Placer Mining in California - Unavailable since 1923, this publication was originally compiled by the California Bureau of Mines. Included are insights into the history of placer gold mining in California, ranging from using a simple gold pan, rocker box or sluice box, right up to the largest of hydraulic mines. All of the major placer gold mining areas are covered in detail, complete with the methods that were used to mine them. This hard to find, previously out of print publication will offer valuable insights for those who are looking for gold and other valuable minerals in California or to those who are interested in mining history. 194 ppgs, 19.99

The Mother Lode Gold System of California - Unavailable since 1929, this publication offers rare insights into the famous Mother Lode Mining Region of California. Included are facts about the local geology, ore deposits, ore genesis and the important gold mines of this important mining area in the California Mother Lode. Includes hard to find details and locations of dozens of hard rock gold mines in the area. This hard to find, previously out of print publication will offer valuable insights for those who are looking for gold and other valuable minerals in California's Mother Lode and surrounding areas, or those who are interested in mining history. 11.99, 132 ppgs

Mines and Minerals of California - Unavailable since 1899, this publication offers rare insights into the early mining industry of California. Included are facts about the early mining history of California, including details on the State's famous gold mining areas, quicksilver mining, copper mining and the early California petroleum industry. This hard to find, previously out of print publication will offer valuable insights for those who are looking for gold or other minerals in California or those who are interested in mining history. 24.99, 458 ppgs

California Golden Treasures: Placer Gold Mining in California in the 1850's - "The Autobiography of Charles Peters: The Good Luck Days of Placer Mining in the 1850's". It was first published in 1915, and later reprinted under the title of "California Golden Treasures", but few copies remain available today. In 1915, Charles Peters was "the oldest pioneer living in California, who mined in ... the days of '49". He was born in Portugal in 1825, first visited California in 1846 as a merchant seaman and returned three years later to seek gold at Columbia, Jackson Creek, and Mokelumne Hill. "California Golden Treasures" is the memoir of his life through the 1850s, followed by a series of "Good Luck" stories, miscellaneous tales of the mining camps, a few of which seem to be credited to Peters, although most were actually the work of another author, drawn from many sources. Also included are many historical happenings, interesting incidents and illustrations of the old mining towns in the good luck era, the placer mining days of the '50s. 19.99, 262 ppgs

Gemstones of California - First published in 1905 as "Gems, Jeweler's Materials and Ornamental Stones of California", it has been unavailable since those days and sheds important light on the gems and precious stones that may be found in California. Included are chapters on diamonds, corundum, topaz, spinel, beryl, garnet, tourmaline, quartz, chalcedony, chrysoprase, jasper, opal, albite, orthoclase, labrodite, jade, lapis lazuli, epidote, apatite, fluorite, hematite, azurite, malachite, turquoise, amber, cat's-eye, pearl, abalone and many others. Included are details on where these gems and precious stones may be found in California, as well as their characteristics. Also included is a chapter on California's gem mines. 15.99, 198 ppgs

Placer Mining For Gold In California - Unavailable since 1946, this publication offers rare insights into the early mining industry of California. Included are facts about the various historical methods of placer mining utilized in California, as well as critical insights into how and where to find placer gold in California. This hard to find, previously out of print publication will offer valuable insights for those who are looking for gold in California, whether they are just starting out or whether they consider themselves an old hand at it. Quite possibly the most informative book available on the subject of placer gold mining. 24.99, 384 ppgs

Butte County, California: Its Advantages and Resources - Mining historian Kerby Jackson introduces us to a classic work of California history in this important re-issue of "Butte County Its Resources and Advantages" which was written by the Rev. Jesse Wood in 1888. This short booklet informs the reader about the early development and resources of Butte County, California. The booklet offers a unique look at Butte County's early communities and their industries in the late 19th century. 12.99, 108 ppgs

What Butte County Offers The Homeseeker - Mining historian Kerby Jackson introduces us to a classic work of California history in this important re-issue of "What Butte County Offers the Homeseeker" which was written by George C. Mansfield and Walter M. Smith in 1919. This short booklet informs the reader about the early development and resources of Butte County, California and the opportunities that were available to those interested in locating to the area just after World War One. The booklet offers a unique look at Butte County's early communities and their industries, namely gold mining and agriculture. 9.99, 68 ppgs

Butte: The Story of a California County - Mining historian Kerby Jackson introduces us to a classic work of California history in this important re-issue of "Butte: The Story of a California County" which was written by George C. Mansfeld in 1919. This hard to find booklet tells the story of how Butte County, California first came into existence, starting with details about its first native inhabitants who lived there before the coming of the white man and his first settlements which was a result of the 1849 California Gold Rush. It goes on to discuss the lives of the first settlers, rounding them out with details about their quest for gold in what became Butte County. Also featured are details on early towns in the county, how they were governed and how their early occupants did their work and spent their leisure time. 9.99, 68 ppgs

The History of Butte County, California: In Two Volumes - Mining historian Kerby Jackson introduces us to a classic work of California history in this important re-issue of "The History of Butte County, California: In Two Volumes" which was written by Harry Laurenz Wells, Frank Gilbert and W.L. Chambers in 1882. This hard to find publication consists of two major parts. The first is a 126 page history of the early settlement of California from 1513 to 1850. The second portion is a history of Butte County from its earliest days of settlement to the early 1880. Chapter topics in the first section include Discovery of and Failure to Occupy California by Spain, Occupation of Lower California by the Jesuits, Conquest of Upper California by the Franciscans, Downfall of the Missions, Spanish Military Occupation, California as a Mexican Territory, The Bear Flag War, The Flores Insurrection, California Admitted to the Union, The Great Fur Companies and their Trapping Expeditions, Settlement of the Sacramento Valley and The Discovery of Gold in California. Chapter topics in the second part, include the Organization of Butte County, Its Early History, Changes in County Boundaries, Formation of Townships, Butte County's County Seat and Courthouse, Butte County Hospital and Infirmary, Elections and County Officers, A History of Crime in Butte County, A History of the Bench and Bar, Press of Butte County, Navigation on the Sacramento River, County Stage Routes, Butte County Agriculture, Butte County's Mining Industry and Indian Difficulties. Also included are details on local communities in Butte County such as Chico City, Oroville, Nord, Anita, Cana, Biggs, Gridley, Nelson, Durham, Dayton, Oregon City, Cherokee, Pence's Ranch, Magalia or Dogtown, Concow Township, Yankee Hill, Concow Valley, Bidwell's Bar, Hamilton, Thompson's Flat, Powellton, Inskip, Lovelocks, Stringtown, Enterprise, Forbestown, Clipper Mills, Bangor, Wyandotte, Boston Ranch or Hurlton and others, many of which are now considered ghost towns.

Also included are insights into the geology of the county and a history of local churches and schools. Also included are the biographies of 42 early settlers in Butte County, Caliornia. This text is heavily illustrated with 50 plus plates depicting important figures in California history, as well as various historic locations in Butte County.   24.99, 366 ppgs

Sights in the Gold Region of Oregon and California - Unavailable since 1853, this publication provides a fascinating insight into the California and Oregon Gold Rushes through the eyes of one of the men who went West and "saw the elephant" to take part in it. Theodore Taylor Johnson's memoir of his journey to the gold fields of California and Oregon offers a unique look into this important time during the settling of the Far West. 382 ppgs, 24.99

Colorado Mining Books

Ores of The Leadville Mining District - Unavailable since 1926, this publication was originally compiled by the United States Department of Interior. This volume also includes important insights into the ores and mineralization of the Leadville Mining District in Colorado. Topics include historic ore prospecting methods, local geology, insights into ore veins and stockworks, the local trend and distribution of ore channels, reverse faults, shattered rock above replacement ore bodies, mineral enrichment in oxidized and sulphide zones and more. 8.5" X 11", 66 ppgs, Retail Price: $8.99

Mining in Colorado - Unavailable since 1926, this publication was originally compiled by the United States Department of Interior. This volume also includes important insights into the mining history of Colorado from its early beginnings in the 1850's right up to the mid 1920's. Not only is Colorado's gold mining heritage included, but also its silver, copper, lead and zinc mining industry. Each mining area is treated separately, detailing the development of Colorado's mines on a county by county basis. 8.5" X 11", 284 ppgs, Retail Price: $19.99

Gold Mining in Gilpin County Colorado - Unavailable since 1876, this publication was originally compiled by the Register Steam Printing House of Central City, Colorado. A rare glimpse at the gold mining history and early mines of Gilpin County, Colorado from their first discovery in the 1850's up to the "flush years" of the mid 1870's. Of particular interest is the history of the discovery of gold in Gilpin County and details about the men who made those first strikes. Special focus is given to the early gold mines and first mining districts of the area, many of which are not detailed in other books on Colorado's gold mining history. 8.5" X 11", 156 ppgs, Retail Price: $12.99

Mining in the Gold Brick Mining District of Colorado - Important insights into the history of the Gold Brick Mining District, as well as its local geography and economic geology. Also included are the histories and locations of historic mines in this important Colorado Mining District, including the Cortland, Carter, Raymond, Gold Links, Sacramento, Bassick, Sandy Hook, Chronicle, Grand Prize, Chloride, Granite Mountain, Lucille, Gray Mountain, Hilltop, Maggie Mitchell, Silver Islet, Revenue, Roosevelt, Carbonate King and others. In addition to hardrock mining, are also included are details on gold placer mining in this portion of Colorado. 8.5" X 11", 140 ppgs, Retail Price: $12.99

Ore Deposits of the London Fault of Colorado - First published in 1941, it has been unavailable since those days and sheds important light on the mines and mineral deposits of the London Fault in Central Colorado's Alma Mining District. This publication sheds important light on the gold veins and lead-silver deposits of the Alma Mining District. Included are geologic details on the London Mine, American Mine, Havigorst Tunnel, Ophir Mine, Mosher Tunnel, London-Butte Mine, Venture Shaft, Hard-To-Beat Mine, Oliver Twist Tunnel, Sacramento Mine, Mudsill Mine, Sherwood Mine, Wagner, Barcoe Tunnel and other mines in this important mining region. 110 ppgs., 10.99

The Mines of Colorado - First published in 1867, it has been unavailable since those days and sheds important light on Colorado's early mining history. Written shortly after the events took place, this publication sheds important light on the Pike's Peak Gold Rush, the discovery of gold on Ralston Creek and Dry Creek in the 1850's, as well as details on the first wave of miners into Colorado and their trials and tribulations as they crossed the Great Plains. Also included are details on early discoveries of lode gold in the mountainous regions of Colorado, details on the early mines hardrock and placer mines, and much more. It is a veritable treasure trove on Colorado's early mining history and will be of great importance to anyone who is interested in the mining of gold or other minerals in Colorado, as well as those interested in the history of the state. 478 ppgs., 29.99

The La Plata Mining District of Colorado - Originally titled "Geology and Ore Deposits in the Vicinity of the La Plata District of Colorado" and first published in 1949, it has been unavailable since those days and sheds important light on the mines and mineral deposits of the La Plata Mining District of Colorado.214 ppgs., 19.99

The Carbonates of Leadville and the Formation of Coal in Colorado - First published in 1879, "Carbonates of Leadville: A Treatise on the Formation of Coal in Colorado" has been unavailable since those days and sheds important light on the history of the famous mining area in Colorado. Featured here are insights into the geology of Carbonates near Leadville, Colorado and the formation of coal deposits in that area. Also included are details on Assaying, Cupellation and Scorification methods used in mining. 112 ppgs., 11.99

The Catalpa Mining Company of Leadville, Colorado - This reprinted circular of the Catalpa Mining Company offers insight into one of the most important mines in the Leadville Mining District during the early 1880's - the famous Catalpa Mine. Also included are details on several adjoining mines in the area including the Evening Star, Pendery, Crescent, Carbonate, Yankee Doodle, Morning Star, Modoc, Etna and others. 70 ppgs., 8.99

The Summitville Mining District of the San Juan Mountains of Colorado - First published in 1960, "Geology and Ore Deposits of the Summitville Mining District San Juan Mountains of Colorado" has been unavailable since those days and sheds important light on the history of the famous mining area in Summit County, Colorado. Featured here in this fascinating text are insights into the local geology of this important Colorado Mining area. Lavishly illustrated with rare photos and hard to find mine maps. 104 ppgs., 10.99

Economic Geology of the Silverton Quadrangle of Colorado - First published in 1901, "Economic Geology of the Silverton Quadrangle Colorado" has been unavailable since those days and sheds important light on the history of the famous mining area in Colorado. Featured here in this fascinating text are insights into the local geology of this important Colorado Mining area, as well as into numerous mines. Lavishly illustrated with rare photos and hard to find mine maps. 304 ppgs., 24.99

Mining in Colorado in 1920 - First published in 1921, "The Annual Report for 1920" by the Colorado Bureau of Mines has been unavailable since those days and sheds important light on the history of the famous mining area in Colorado. Featured here are insights into the mineral industry of Colorado as it existed in 1920, complete with full statistics of all known operating gold, silver, copper and other mines that operated throughout the state. 90 ppgs., 9.99

Geology of the Glenwood Springs Quadrangle of North Western Colorado - First published in 1963, "Geology of the Glenwood Springs Quadrangle and Vicinity of North Western Colorado" has been unavailable since those days and sheds important light on the geology of this portion of North Western Colorado. Included are details on dozens of mines in this important mining Colorado region. 102 ppgs., 11.99

Men of Note Affiliated With Mining in the Cripple Creek Mining District - First published in 1905 by L.A. Snyder, "Men of note affiliated with mining and mining interests in the Cripple Creek district " has been unavailable since those days and sheds important light on the history of the famous Cripple Creek Mining District. Featured here in this fascinating text are insights into the movers and shakers who put the Cripple Creek Mining District on the map back in its heyday, including men like Bob Womack, Frank Campbell, James Wright, George Hill, J.E. Jones, Walter Swanson and dozens of others who discovered, owned and managed the leading mines of the Cripple Creek District. Also included are rare insights into the early mining history of this important mining district. Lavishly illustrated with rare photos from the early days of mining in Cripple Creek. 156 ppgs., 14.99

Ore Deposits Near Lake City, Colorado - First published in 1911, "Geology and Ore Deposits Near Lake City, Colorado" has been unavailable since those days and sheds important light on the geology and mining areas of this portion Colorado. Included are details on dozens of mines in this important mining Colorado region. 164 ppgs., 14.99

Ore Deposits of the Montezuma Quadrangle of Colorado - First published in 1935, "Ore Deposits of the Montezuma Quadrangle of Colorado" has been unavailable since those days and sheds important light on the history of the famous mining area in Summit County, Colorado. Featured here in this fascinating text are insights into the local geology, as well as dozens of important gold, silver and copper mines. Lavishly illustrated with rare photos and hard to find mine maps. 190 ppgs., 19.99

Ore Deposits of the Platoro and Summitville Mining Districts of Colorado - First published in 1917, "Ore Deposits of the Platoro and Summitville Mining Districts of Colorado" has been unavailable since those days and sheds important light on the history of the famous mining area in Summit County, Colorado. Featured here in this fascinating text are insights into the local geology, as well as dozens of mines in the Summitville, Platoro, Gilmore, Stunner and Jasper Mining Districts, including the Eurydice, Pass-Me-By, Asiatic, Watrous, Perry, Miser, Guadaloupe, Forest King, Parole, Morrimac, Congress, Mammoth, Golconda, Little Annie, Bobtail, Aztec and others. Lavishly illustrated with rare photos and hard to find mine maps. 184 ppgs., 19.99

Ore Deposits of the Creede Mining District of Colorado - First published in 1923, it has been unavailable since those days and sheds important light on the mines and mineral deposits of the Creede Mining District in Mineral County, Colorado. Included are geologic details on dozens of mines in this important mining Colorado region. 242 ppgs., 19.99

Ore Deposits of the Garfield Quadrangle of Colorado - First published in 1957, it has been unavailable since those days and sheds important light on the mines and mineral deposits of the Garfield Quadrangle in Garfield County, Colorado. Included are geologic details on dozens of mines in this important mining Colorado region. 144 ppgs., 12.99

Ore Deposits of the Breckenridge Mining District of Colorado - First published in 1911, it has been unavailable since those days and sheds important light on the mines and mineral deposits of the Breckenridge Mining District in Summit County, Colorado. Included are geologic details on dozens of mines in this important mining Colorado region.   246 ppgs., 19.99

Ore Deposits of the Bonanza Mining District of Colorado - First published in 1932, it has been unavailable since those days and sheds important light on the mines and mineral deposits of the Bonanza Mining District. Included are geologic details on dozens of mines in this important mining Colorado region.   232 ppgs., 24.99

East Coast Mining Books

The Gold Fields of the Southern Appalachians - Unavailable since 1895, this important publication was originally published by the US Department of Interior and has been unavailable for nearly 120 years. Topics include the geology, rock formations and the formation of ore deposits in this important mining area of the American South. Of particular focus is information on the history and statistics of the ore deposits in this area, their form and structure and veins. Also included are details on the placer gold deposits of the region. The gold fields of the Georgian Belt, Carolinian Belt and the South Mountain Mining District of North Carolina are all treated in descriptive detail. Included are hard to find details, including the descriptions and locations of numerous gold mines in Georgia, North Carolina and elsewhere in the American South. Also included are details on the gold belts of the British Maritime Provinces and the Green Mountains. **8.5" X 11", 104 ppgs, Retail Price: $9.99**

Gold Rush Tales Series

Millions in Siskiyou County Gold - In this first volume of the "Gold Rush Tales" series, leading mining historian and editor Kerby Jackson, introduces us to the story of how millions of dollars worth of gold was discovered in Siskiyou County during the California Gold Rush. Lavishly illustrated with photos from the 19th Century, this hard to find information was first published in 1897 and sheds important light onto the gold rush era in Siskiyou County, California and the experiences of the men who dug for the gold and actually found it. **8.5" X 11", 82 ppgs, Retail Price: $9.99**

The California Rand in the Days of '49 - In this second volume of the "Gold Rush Tales" series, leading mining historian and editor Kerby Jackson, introduces us to four tales from the California Gold Rush. Lavishly illustrated with photos from the 19th Century, this hard to find information was first published in 1890's and includes the stories of "California's Rand", details about Chinese miners, how one early miner named Baker struck it rich and also the story of Alphonzo Bowers, who invented the first hydraulic gold dredge. **8.5" X 11", 54 ppgs, Retail Price: $9.99**

Idaho Mining Books

Gold in Idaho - Unavailable since the 1940's, this publication was originally compiled by the Idaho Bureau of Mines and includes details on gold mining in Idaho. Included is not only raw data on gold production in Idaho, but also valuable insight into where gold may be found in Idaho, as well as practical information on the gold bearing rocks and other geological features that will assist those looking for placer and lode gold in the State of Idaho. This volume also includes thirteen gold maps that greatly enhance the practical usability of the information contained in this small book detailing where to find gold in Idaho. **8.5" X 11", 72 ppgs. Retail Price: $9.99**

Geology of the Couer D'Alene Mining District of Idaho - Unavailable since 1961, this publication was originally compiled by the Idaho Bureau of Mines and Geology and includes details on the mining of gold, silver and other minerals in the famous Coeur D'Alene Mining District in Northern Idaho. Included are details on the early history of the Coeur D'Alene Mining District, local tectonic settings, ore deposit features, information on the mineral belts of the Osburn Fault, as well as detailed information on the famous Bunker Hill Mine, the Dayrock Mine, Galena Mine, Lucky Friday Mine and the infamous Sunshine Mine. This volume also includes sixteen hard to find maps. **8.5" X 11", 70 ppgs. Retail Price: $9.99**

The Gold Camps and Silver Cities of Idaho - From 1963, this important publication on Idaho Mining has not been available for nearly fifty years. Included are rare insights into the history of Idaho's Gold Rush, as well as the mad craze for silver in the Idaho Panhandle. Documented in fine detail are the early mining excitements at Boise Basin, at South Boise, in the Owyhees, at Deadwood, Long Valley, Stanley Basin and Robinson Bar, at Atlanta, on the famous Boise River, Volcano, Little Smokey, Banner, Boise Ridge, Hailey, Leesburg, Lemhi, Pearl, at South Mountain, Shoup and Ulysses, Yellow Jacket and Loon Creek. The story follows with the appearance of Chinese miners at the new mining camps on the Snake River, Black Pine, Yankee Fork, Bay Horse, Clayton, Heath, Seven Devils, Gibbonsville, Vienna and Sawtooth City. Also included are special sections on the Idaho Lead and Silver mines of the late 1800's, as well as the mining discoveries of the early 1900's that paved the way for Idaho's modern mining and mineral industry. Lavishly illustrated with rare historic photos, this volume provides a one of a kind documentary into Idaho's mining history that is sure to be enjoyed by not only modern miners and prospectors who still scour the hills in search of nature's treasures, but also those enjoy history and tromping through overgrown ghost towns and long abandoned mining camps. **186 ppgs, $14.99**

Ore Deposits and Mining in North Western Custer County Idaho - Unavailable since 1913, this important publication was originally published by the Us Department of the Interior and has been unavailable for a century. Included are fine details on the geology, geography, gold placers and gold and silver bearing quartz veins of the mining region of North West Custer County, Idaho. Of particular interest is a rare look at the mines and prospects of the region, including those such as the Ramshorn Mine, SkyLark, Riverview, Excelsior, Beardsley, Pacific, Hoosier, Silver Brick, Forest Rose and dozens of others in the Bay Horse Mining District. Also covered are the mines of the Yankee Fork District such as the Lucky Boy, Badger, Black, Enterprise, Charles Dickens, Morrison, Golden Sunbeam, Montana, Golden Gate and others, as well as those in the Loon Mining District. **8.5" X 11", 126 ppgs. Retail Price: $12.99**

Gold Rush To Idaho - Unavailable since 1963, this important publication was originally published by the Idaho Bureau of Mines and has been unavailable for 50 years. "Gold Rush To Idaho" revisits the earliest years of the discovery of gold in Idaho Territory and introduces us to the conditions that the pioneer gold seekers met when they blazed a trail through the wilderness of Idaho's mountains and discovered the precious yellow metal at Oro Fino and Pierce. Subsequent rushes followed at places like Elk City, Newsome, Clearwater Station, Florence, Warrens and elsewhere. Of particular interest is a rare look at the hardships that the first miners in Idaho met with during their day to day existences and their attempts to bring law and order to their mining camps. **8.5" X 11", 88 ppgs. Retail Price: $9.99**

The Geology and Mines of Northern Idaho and North Western Montana - Unavailable since 1909, this important publication was originally published by the Us Department of the Interior and has been unavailable for a century. Included are fine details on the geology and geography of the mining regions of Northern Idaho and North Western Montana. Of particular interest is a rare look at the mines and prospects of the region, including those in the Pine Creek Mining District, Lake Pend Oreille district, Troy Mining District, Sylvanite District, Cabinet Mining District, Prospect Mining District and the Missoula Valley. Some of the mines featured include the Iron Mountain, Silver Butte, Snowshoe, Grouse Mountain Mine and others. **8.5" X 11", 142 ppgs. Retail Price: $12.99**

Mining in the Alturas Quadrangle of Blaine County Idaho - Unavailable since 1922, this important publication was originally published by the Idaho Bureau of Mines and has been unavailable for ninety years. Topics include the geology, rock formations and the formation of ore deposits in this important mining area of Idaho. Of particular focus is information on the local geology, quartz veins and ore deposits of this portion of Idaho. Included are hard to find details, including the descriptions and locations of numerous gold and silver mines in the area including the Silver King, Pilgrim, Columbia, Lone Jack, Sunbeam, Pride of the West, Lucky Boy, Scotia, Atlanta, Beaver-Bidwell and others mines and prospects. **8.5" X 11", 56 ppgs. Retail Price: $8.99**

Mining in Lemhi County Idaho - Originally published in 1913, this important book on Idaho Mining has not been available to miners for over a century. Included are rare insights into hundreds of gold, silver, copper and other mines in this famous Idaho mining area. Details include the locations, geology, history, production and other facts of the mines of this region, not only gold and silver hardrock mines, but also gold placer mines, lead-silver deposits, copper mines, cobalt-nickel deposits, tungsten and tin mines . It is lavishly illustrated with hard to find photos of the period and rare mining maps. Some of the vicinities featured include the Nicholia Mining District, Spring Mountain District, Texas District, Blue Wing District, Junction District, McDevitt District, Pratt Creek, Eldorado District, Kirtley Creek, Carmen Creek, Gibbonsville, Indian Creek, Mineral Hill District, Mackinaw, Eureka District, Blackbird District, YellowJacket District, Gravel Range District, Junction District, Parker Mountain and other mining districts. **8.5" X 11", 226 ppgs. Retail Price: $19.99**

Mining in Shoshone County Idaho - First published in 1923, it has been unavailable for over a century and sheds important light on the mining history of Shoshone County, Idaho. Some of the topics include the history of mining in Shoshone County, a look at the local geology and ore characteristics of lead-silver deposits, zinc deposits, copper, antimony, gold and other minerals. Also included are insights into the history, production, characteristics and locations of numerous mines in the area. 198 ppgs, 15.99

Geology of the Bitterroot and Clearwater Mountains of Idaho and Montana - Unavailable since 1904, this publication offers rare insights into the geology of this region of Idaho and Montana. Included are also details on the numerous gold, silver and copper mines of this region. 13.99, 150 ppgs

Gold in the Black Pine Mining District of Idaho - Unavailable since 1984, this publication offers rare insights into the famous Black Pine Mining District of Idaho. Included in this very small booklet are facts about the geology and ore deposits of this famous mining district in Idaho, as well as some insight into the early mining history of the district. 6.99, 44 pgs

Mining in Eastern Cassia County Idaho - Unavailable since 1931, this publication offers rare insights into this famous mining region of Idaho. Included are descriptions of numerous gold and silver mines, their locations, how they were established and how they operated, as well as their geologic structures. 19.99, 226 ppgs

<u>Geology of the Alder Creek Mining District of Idaho</u> - Unavailable since 1968, this publication offers rare insights into the famous Alder Creek Mining District of Idaho. Included in this small booklet are facts about the geology and ore deposits of this famous mining district in Idaho. 7.99, 58 ppgs

<u>Mines of the Alder Creek Mining District of Idaho</u> - Unavailable since 1997, this publication offers rare insights into the famous Alder Creek Mining District of Idaho. Included in this small booklet are facts about Empire Mine, Blue Bird or Easlie Mine, Champion Mine, Doughboy Mine, Horseshoe Mine and White Knob Group. Included are descriptions of each mine, their locations, how they were established and how they operated. Lavishly illustrated with hard to find mine maps and rare historical photographs. 8.99, 80 ppgs

Montana Mining Books

<u>A History of Butte Montana: The World's Greatest Mining Camp</u> - First published in 1900 by H.C. Freeman, this important publication sheds a bright light on one of the most important mining areas in the history of The West. Together with his insights, as well as rare photographs of the periods, Harry Freeman describes Butte and its vicinity from its early beginnings, right up to its flush years when copper flowed from its mines like a river. At the time of publication, Butte, Montana was known worldwide as "The Richest Mining Spot On Earth" and produced not only vast amounts of copper, but also silver, gold and other metals from its mines. Freeman illustrates, with great detail, the most important mines in the vicinity of Butte, providing rare details on their owners, their history and most importantly, how the mines operated and how their treasures were extracted. Of particular interest are the dozens of rare photographs that depict mines such as the famous Anaconda, the Silver Bow, the Smoke House, Moose, Paulin, Buffalo, Little Minah, the Mountain Consolidated, West Greyrock, Cora, the Green Mountain, Diamond, Bell, Parnell, the Neversweat, Nipper, Original and many others.   8.5″ X 11″, 142 ppgs. Retail Price: $12.99

<u>The Butte Mining District of Montana</u> - This important publication on Montana Mining has not been available for over a century. Included are rare insights into the gold, copper and silver mines of Butte, Montana together with hard to find maps and photographs. Some of the topics include the early history of gold, silver and copper mining in the Butte area, insight into the geology of its mining areas, the local distribution of gold, silver and copper ores, as well their composition and how to identify them. Also included are detailed facts about the mines in the Butte Mining District, including the famous Anaconda Mine, Gagnon, Parrot, Blue Vein, Moscow, Poulin, Stella, Buffalo, Green Mountain, Wake Up Jim, the Diamond-Bell Group, Mountain Consolidated, East Greyrock, West Greyrock, Snowball, Corra, Speculator, Adirondack, Miners Union, the Jessie-Edith May Group, Otisco, Iduna, Colorado, Lizzie, Cambers, Anderson, Hesperus, Preferencia and dozens of others. 8.5″ X 11″, 298 ppgs. Retail Price: $24.99

<u>Mines of the Helena Mining Region of Montana</u> - This important publication on Montana Mining has not been available for over a century. Included are rare insights into the gold, copper and silver mines of the vicinity of Helena, Montana, including the Marysville Mining District, Elliston Mining District, Rimini Mining District, Helena Mining District, Clancy Mining District, Wickes Mining District, Boulder and Basin Mining Districts and the Elkhorn Mining District. Some of the topics include the early history of gold, silver and copper mining in the Helena area, insight into the geology of its mining areas, the local distribution of gold, silver and copper ores, as well their composition and how to identify them. Also included are detailed facts, history, geology and locations of over one hundred gold, silver and copper mines in the area . 8.5″ X 11″, 162 ppgs, Retail Price: $14.99

<u>Mines and Geology of the Garnet Range of Montana</u> - This important publication on Montana Mining has not been available for over a century. Included are rare insights into the gold, copper and silver mines of the vicinity of this important mining area of Montana. Some of the topics include the early history of gold, silver and copper mining in the Garnet Mountains, insight into the geology of its mining areas, the local distribution of gold, silver and copper ores, as well their composition and how to identify them. Also included are detailed facts, history, geology and locations of numerous gold, silver and copper mines in the area . 8.5″ X 11″, 100 ppgs, Retail Price: $11.99

<u>Mines and Geology of the Philipsburg Quadrangle of Montana</u> - This important publication on Montana Mining has not been available for over a century. Included are rare insights into the gold, copper and silver mines of the vicinity of this important mining area of Montana. Some of the topics include the early history of gold, silver and copper mining in the Philipsburg Quadrangle, insight into the geology of its mining areas, the local distribution of gold, silver and copper ores, as well their composition and how to identify them. Also included are detailed facts, history, geology and locations of over one hundred gold, silver and copper mines in the area 8.5″ X 11″, 290 ppgs, Retail Price: $24.99

<u>Geology of the Marysville Mining District of Montana</u> - Included are rare insights into the mining geology of the Marysville Mining District. Some of the topics include the early history of gold, silver and copper mining in the area, insight into the geology of its mining areas, the local distribution of gold, silver and copper ores, as well their composition and how to identify them. Also included are detailed facts, history, geology and locations of gold, silver and copper mines in the area   8.5″ X 11″, 198 ppgs, Retail Price: $19.99

<u>The Geology and Mines of Northern Idaho and North Western Montana</u>- See listing under Idaho.

<u>The History of Gold Dredging in Montana</u> - Unavailable since 1916, this important publication was originally published by the Us Bureau of Mines and has been unavailable for a century. A century and more ago, giant dredging machines dug in Montana's rivers and creeks in search of illusive golden riches. First appearing in California in the 1850's, gold dredges finally reached their peak of development in Siberia and New Zealand before becoming popular again in the United States. This book offers a unique historical perspective on the gold dredges that once operated in Montana. This book on Montana mining history is lavishly illustrated with dozens of rare historic photos gold dredges that once operated in Montana, as well as hard to locate plans on how these dredges were designed. 120 ppgs., 11.99

The Great Dynamite Explosion at Butte, Montana - On the night of January 15th, 1895, the great mining center of Butte, Montana was devastated by a series of explosions. As the Reno Daily Journal's headline blared: DYNAMITE EXPLOSION. Terrible Loss of Life at Butte, Montana. ABOUT 150 KILLED AND INJURED. The Fire Department Annihilated-Windows Demolished a Mile Away. The Daily Journal continued, "A fire broke out in the Butte Hardware Company's warehouse near Butte City, Montana. There was a large quantity of giant powder stored in the building and when the Fire Department was fighting the flames the powder exploded killing every fireman except two. While the dead and wounded were being removed another explosion occurred which killed more persons, including several policemen and citizens. Many persons were torn to fragments and others were shocked to death by the concussion. Later a third explosion occurred increasing the number of deaths and adding to the ruin and devastation." Almost as soon as the fires had cooled, local educator John F. Davies set pen to paper to record for history what took place, including the accounts of some of those who saw history unfold first hand. 74 ppgs., 9.99

Nevada Mining Books

The Bull Frog Mining District of Nevada - Unavailable since 1910, this publication was originally compiled by the United States Department of Interior. This volume also includes important insights into the geologic formations, faults and other aspects of economic geology in this Nevada mining district. Of particular interest are the fine details on many mines in the area, including their locations, histories, development and mineralization. Some of the mines featured include the National Bank Mine, Providence, Gibraltor, Tramps, Denver, Original Bullfrog, Gold Bar, Mayflower, Homestake-King and other mines and prospects. **8.5" X 11", 152 ppgs, Retail Price: $14.99**

<u>History of the Comstock Lode</u> - Unavailable since 1876, this publication was originally released by John Wiley & Sons. This volume also includes important insights into the famous Comstock Lode of Nevada that represented the first major silver discovery in the United States. During its spectacular run, the Comstock produced over 192 million ounces of silver and 8.2 million ounces of gold. Not only did the Comstock result in one of the largest mining rushes in history and yield immense fortunes for its owners, but it made important contributions to the development of the State of Nevada, as well as neighboring California. Included here are important details on not only the early development and history of the Comstock, but also rare early insight into its mines, ore and its geology.**8.5" X 11", 244 ppgs, Retail Price: $19.99**

The Pioche Mining District of Nevada - First published in 1932, it has been unavailable for over a century and sheds important light on the mining history of Nevada. Some of the topics include the history of mining in this district, as well as the characteristics of its mineral and ore deposits. Also included are insights into the history, production, characteristics and locations of numerous mines in the area. Some of the mines include the Combined Metals, Pioche, Ely Valley, No. 10, Poorman, Wide Awake, Alps, Prince, Virginia Louise, Half Moon, Abe Lincoln, Fairview, Bristol Silver, National, Vesuvius, Inman, Tempest, Hillside, Jackrabbit, Lucky Star, Fortuna, Mendha, Manhattan, Hamburg, Comet, Lyndon and others. 108 ppgs 10.99

The Yerington Mining District of Nevada - First published in 1932, it has been unavailable for over a century and sheds important light on the mining history of Nevada. Some of the topics include the history of mining in this district, as well as the characteristics of its mineral and ore deposits. Also included are insights into the history, production, characteristics and locations of numerous mines in the area. Some of the mines include the Bluestone, Mason Valley, Malachite, McConnell, Greenwood, Western Nevada, Ludwig, Douglas Hill, Casting Copper, Montana-Yerington, Empire, Jim Beatty, Terry and McFarland, Blue Jay and others. 92 ppgs, 10.99

The Genesis of the Ores of Tonopah Nevada - Unavailable since 1918, this hard to find publication includes valuable insights into the gold mines around Tonopah, Nevada. The publication includes important details into the geology of mines in the Tonopah Mining District of Nevada. 90 ppgs, 10.99

Mining Camps of Elko, Lander and Eureka Counties Nevada - Unavailable since 1910, this hard to find publication includes valuable insights into the mining camps of Elko, Lander and Eureka Counties, Nevada. The publication includes important details into the history of mines and mining in these three Nevada counties. 154 ppgs, 12.99

<u>Ore Deposits of the Bullfrog Quadrangle</u> - Unavailable since 1964 and released as "Geology of Bullfrog Quadrangle and Ore Deposits Related to Bullfrog Hills Caldera, Nye County, Nevada and Inyo County, California". The publication includes important details into the geology of mines in the Bullfrog Quadrangle of Nye County, Nevada and Inyo County, California. 52 ppgs, 9.99

<u>Mining in Eureka County Nevada</u> - Unavailable since 1879, this hard to find publication includes valuable insights into the early mining history off Eureka County, Nevada. The publication includes important details into the early history of the mines of Eureka County, as well as their development, production and how their ores were treated. Also included are details on the 1872 Mining Act, as well as the local rules, regulations and customs of the miners in Eureka County.134 ppgs, 12.99

New Mexico Mining Books

<u>The Mogollon Mining District of New Mexico</u> - Unavailable since 1927, this important publication was originally published by the US Department of Interior and has been unavailable for 80 years. Topics include the geology, rock formations and the formation of ore deposits in this important mining area in New Mexico. Of particular focus is information on the history and production of the ore deposits in this area, their form and structure, vein filling, their paragenesis, origins and ore shoots, as well as oxidation and supergene enrichment. Also included are hard to find details, including the descriptions and locations of numerous gold, silver and other types of mines, including the Eureka, Pacific, South Alpine, Great Western, Enterprise, Buffalo, Mountain View, Floride, Gold Dust, Last Chance, Deadwood, Confidence, Maud S., Deep Down, Little Fanney, Trilby, Johnson, Alberta, Comet, Golden Eagle, Cooney, Queen, the Iron Crown, Eberle, Clifton, Andrew Jackson mine, Mascot and others. **8.5" X 11", 144 ppgs, Retail Price: $12.99**

<u>The Percha Mining District of Kingston New Mexico</u> - Unavailable since 1883, this important publication was originally published by the Kingston Tribune and has been unavailable for over one hundred and thirty five years. Having been written during the earliest years of gold and silver mining in the Percha Mining District, unlike other books on the subject, this work offers the unique perspective of having actually been written while the early mining history of this area was still being made. In fact, the work was written so early in the development of this area that many of the notable mines in the Percha District were less than a few years old and were still being operated by their original discoverers with the same enthusiasm as when they were first located. Included are hard to find details on the very earliest gold and silver mines of this important mining district near Kingston in Sierra County, New Mexico. **8.5" X 11", 68 ppgs, Retail Price: $9.99**

<u>Economic Geology of New Mexico</u> - Written in 1908, this hard to find publication includes valuable insights into the mining industry of New Mexico. Included are important details on the economic geology of New Mexico, including the general locations of numerous valuable minerals in New Mexico **8.5" X 11", 76 ppgs, Retail Price: $8.99**

<u>The Magdalena Mining District of New Mexico</u> - Written in 1942, this hard to find publication includes valuable insights into the gold and silver mining industry of New Mexico. Included are important details on the geology and ore minerals of the Magdalena Mining District, as well as the locations and other facts of the important gold and silver mines of the area. Some of the mines featured include the Nitt, Graphic-Waldo or Ozark Mine, Kelly Mine, Juanita, South Juanita, Black Cloud, Mistletoe, Young America, Imperial, Enterprise, Linchburg Tunnel, Woodland, Cavern, Grand Ledge, Connelly, West Virginia, Victor, Sampson, Grand Tower, Legal Tender, Germany, Little Loella, Tip Top, Key, Stonewall, Ambrosia, Sleeper, Hardscrabble, Anchor, Vindicator, Cavalier, Heister and others. Lavishly illustrated with photographs of local ore, mine maps and more. **8.5" X 11", 230 ppgs, Retail Price: $19.99**

<u>Mineral Belts of Western Sierra County New Mexico</u> - Written in 1979, this hard to find publication includes valuable insights into the mining industry of New Mexico. Included are important details on the gold and silver bearing mineral belts of Western Sierra County, New Mexico. **8.5" X 11", 80 ppgs, Retail Price: $8.99**

Oregon Mining Books

<u>Geology and Mineral Resources of Josephine County, Oregon</u> - Unavailable since the 1970's, this important publication was originally compiled by the Oregon Department of Geology and Mineral Industries and includes important details on the economic geology and mineral resources of this important mining area in South Western Oregon. Included are notes on the history, geology and development of important mines, as well as insights into the mining of gold, copper, nickel, limestone, chromium and other minerals found in large quantities in Josephine County, Oregon. **8.5" X 11", 54 ppgs. Retail Price: $9.99**

Mines and Prospects of the Mount Reuben Mining District - Unavailable since 1947, this important publication was originally compiled by geologist Elton Youngberg of the Oregon Department of Geology and Mineral Industries and includes detailed descriptions, histories and the geology of the Mount Reuben Mining District in Josephine County, Oregon. Included are notes on the history, geology, development and assay statistics, as well as underground maps of all the major mines and prospects in the vicinity of this much neglected mining district. **8.5" X 11", 48 ppgs. Retail Price: $9.99**

The Granite Mining District - Notes on the history, geology and development of important mines in the well known Granite Mining District which is located in Grant County, Oregon. Some of the mines discussed include the Ajax, Blue Ribbon, Buffalo, Continental, Cougar-Independence, Magnolia, New York, Standard and the Tillicum. Also included are many rare maps pertaining to the mines in the area. **8.5" X 11", 48 ppgs. Retail Price: $9.99**

Ore Deposits of the Takilma and Waldo Mining Districts of Josephine County, Oregon - The Waldo and Takilma mining districts are most notable for the fact that the earliest large scale mining of placer gold and copper in Oregon took place in these two areas. Included are details about some of the earliest large gold mines in the state such as the Llano de Oro, High Gravel, Cameron, Platerica, Deep Gravel and others, as well as copper mines such as the famous Queen of Bronze mine, the Waldo, Lily and Cowboy mines. This volume also includes six maps and 20 original illustrations. **8.5" X 11", 74 ppgs. Retail Price: $9.99**

Metal Mines of Douglas, Coos and Curry Counties, Oregon - Oregon mining historian Kerby Jackson introduces us to a classic work on Oregon's mining history in this important re-issue of Bulletin 14C Volume 1, otherwise known as the Douglas, Coos & Curry Counties, Oregon Metal Mines Handbook. Unavailable since 1940, this important publication was originally compiled by the Oregon Department of Geology and Mineral Industries includes detailed descriptions, histories and the geology of over 250 metallic mineral mines and prospects in this rugged area of South West Oregon. **8.5" X 11", 158 ppgs. Retail Price: $19.99**

Metal Mines of Jackson County, Oregon - Unavailable since 1943, this important publication was originally compiled by the Oregon Department of Geology and Mineral Industries includes detailed descriptions, histories and the geology of over 450 metallic mineral mines and prospects in Jackson County, Oregon. Included are such famous gold mining areas as Gold Hill, Jacksonville, Sterling and the Upper Applegate. **8.5" X 11", 220 ppgs. Retail Price: $24.99**

Metal Mines of Josephine County, Oregon - Oregon mining historian Kerby Jackson introduces us to a classic work on Oregon's mining history in this important re-issue of Bulletin 14C, otherwise known as the Josephine County, Oregon Metal Mines Handbook. Unavailable since 1952, this important publication was originally compiled by the Oregon Department of Geology and Mineral Industries includes detailed descriptions, histories and the geology of over 500 metallic mineral mines and prospects in Josephine County, Oregon. **8.5" X 11", 250 ppgs. Retail Price: $24.99**

Metal Mines of North East Oregon - Oregon mining historian Kerby Jackson introduces us to a classic work on Oregon's mining history in this important re-issue of Bulletin 14A and 14B, otherwise known as the North East Oregon Metal Mines Handbook. Unavailable since 1941, this important publication was originally compiled by the Oregon Department of Geology and Mineral Industries and includes detailed descriptions, histories and the geology of over 750 metallic mineral mines and prospects in North Eastern Oregon. **8.5" X 11", 310 ppgs. Retail Price: $29.99**

Metal Mines of North West Oregon - Oregon mining historian Kerby Jackson introduces us to a classic work on Oregon's mining history in this important re-issue of Bulletin 14D, otherwise known as the North West Oregon Metal Mines Handbook. Unavailable since 1951, this important publication was originally compiled by the Oregon Department of Geology and Mineral Industries and includes detailed descriptions, histories and the geology of over 250 metallic mineral mines and prospects in North Western Oregon. **8.5" X 11", 182 ppgs. Retail Price: $19.99**

Mines and Prospects of Oregon - Mining historian Kerby Jackson introduces us to a classic mining work by the Oregon Bureau of Mines in this important re-issue of The Handbook of Mines and Prospects of Oregon. Unavailable since 1916, this publication includes important insights into hundreds of gold, silver, copper, coal, limestone and other mines that operated in the State of Oregon around the turn of the 19th Century. Included are not only geological details on early mines throughout Oregon, but also insights into their history, production, locations and in some cases, also included are rare maps of their underground workings. **8.5" X 11", 314 ppgs. Retail Price: $24.99**

Lode Gold of the Klamath Mountains of Northern California and South West Oregon
(See California Mining Books)

Mineral Resources of South West Oregon - Unavailable since 1914, this publication includes important insights into dozens of mines that once operated in South West Oregon, including the famous gold fields of Josephine and Jackson Counties, as well as the Coal Mines of Coos County. Included are not only geological details on early mines throughout South West Oregon, but also insights into their history, production and locations. **8.5" X 11", 154 ppgs. Retail Price: $11.99**

Chromite Mining in The Klamath Mountains of California and Oregon
(See California Mining Books)

Southern Oregon Mineral Wealth - Unavailable since 1904, this rare publication provides a unique snapshot into the mines that were operating in the area at the time. Included are not only geological details on early mines throughout South West Oregon, but also insights into their history, production and locations. Some of the mining areas include Grave Creek, Greenback, Wolf Creek, Jump Off Joe Creek, Granite Hill, Galice, Mount Reuben, Gold Hill, Galls Creek, Kane Creek, Sardine Creek, Birdseye Creek, Evans Creek, Foots Creek, Jacksonville, Ashland, the Applegate River, Waldo, Kerby and the Illinois River, Althouse and Sucker Creek, as well as insights into local copper mining and other topics. **8.5" X 11", 64 ppgs. Retail Price: $8.99**

Geology and Ore Deposits of the Takilma and Waldo Mining Districts - Unavailable since the 1933, this publication was originally compiled by the United States Geological Survey and includes details on gold and copper mining in the Takilma and Waldo Districts of Josephine County, Oregon. The Waldo and Takilma mining districts are most notable for the fact that the earliest large scale mining of placer gold and copper in Oregon took place in these two areas. Included in this report are details about some of the earliest large gold mines in the state such as the Llano de Oro, High Gravel, Cameron, Platerica, Deep Gravel and others, as well as copper mines such as the famous Queen of Bronze mine, the Waldo, Lily and Cowboy mines. In addition to geological examinations, insights are also provided into the production, day to day operations and early histories of these mines, as well as calculations of known mineral reserves in the area. This volume also includes six maps and 20 original illustrations. **8.5" X 11", 74 ppgs. Retail Price: $9.99**

Gold Mines of Oregon - Oregon mining historian Kerby Jackson introduces us to a classic work on Oregon's mining history in this important re-issue of Bulletin 61, otherwise known as "Gold and Silver In Oregon". Unavailable since 1968, this important publication was originally compiled by geologists Howard C. Brooks and Len Ramp of the Oregon Department of Geology and Mineral Industries and includes detailed descriptions, histories and the geology of over 450 gold mines Oregon. Included are notes on the history, geology and gold production statistics of all the major mining areas in Oregon including the Klamath Mountains, the Blue Mountains and the North Cascades. While gold is where you find it, as every miner knows, the path to success is to prospect for gold where it was previously found. **8.5" X 11", 344 ppgs. Retail Price: $24.99**

Mines and Mineral Resources of Curry County Oregon - Originally published in 1916, this important publication on Oregon Mining has not been available for nearly a century. Included are rare insights into the history, production and locations of dozens of gold mines in Curry County, Oregon, as well as detailed information on important Oregon mining districts in that area such as those at Agness, Bald Face Creek, Mule Creek, Boulder Creek, China Diggings, Collier Creek, Elk River, Gold Beach, Rock Creek, Sixes River and elsewhere. Particular attention is especially paid to the famous beach gold deposits of this portion of the Oregon Coast. **8.5" X 11", 140 ppgs. Retail Price: $11.99**

Chromite Mining in South West Oregon - Originally published in 1961, this important publication on Oregon Mining has not been available for nearly a century. Included are rare insights into the history, production and locations of nearly 300 chromite mines in South Western Oregon. **8.5" X 11", 184 ppgs. Retail Price: $14.99**

Mineral Resources of Douglas County Oregon - Originally published in 1972, this important publication on Oregon Mining has not been available for nearly forty years. Included are rare insights into the geology, history, production and locations of numerous gold mines and other mining properties in Douglas County, Oregon. **8.5" X 11", 124 ppgs. Retail Price: $11.99**

Mineral Resources of Coos County Oregon - Originally published in 1972, this important publication on Oregon Mining has not been available for nearly forty years. Included are rare insights into the geology, history, production and locations of numerous gold mines and other mining properties in Coos County, Oregon. **8.5" X 11", 100 ppgs. Retail Price: $11.99**

Mineral Resources of Lane County Oregon - Originally published in 1938, this important publication on Oregon Mining has not been available for nearly seventy five years. Included are extremely rare insights into the geology and mines of Lane County, Oregon, in particular in the Bohemia, Blue River, Oakridge, Black Butte and Winberry Mining Districts. **8.5" X 11", 82 ppgs. Retail Price: $9.99**

Mineral Resources of the Upper Chetco River of Oregon: Including the Kalmiopsis Wilderness - Originally published in 1975, this important publication on Oregon Mining has not been available for nearly forty years. Withdrawn under the 1872 Mining Act since 1984, real insight into the minerals resources and mines of the Upper Chetco River has long been unavailable due to the remoteness of the area. Despite this, the decades of battle between property owners and environmental extremists over the last private mining inholding in the area has continued to pique the interest of those interested in mining and other forms of natural resource use. Gold mining began in the area in the 1850's and has a rich history in this geographic area, even if the facts surrounding it are little known. Included are twenty two rare photographs, as well as insights into the Becca and Morning Mine, the Emmly Mine (also known as Emily Camp), the Frazier Mine, the Golden Dream or Higgins Mine, Hustis Mine, Peck Mine and others. **8.5" X 11", 64 ppgs. Retail Price: $8.99**

Gold Dredging in Oregon - Originally published in 1939, this important publication on Oregon Mining has not been available for nearly seventy five years. Included are extremely rare insights into the history and day to day operations of the dragline and bucketline gold dredges that once worked the placer gold fields of South West and North East Oregon in decades gone by. Also included are details into the areas that were worked by gold dredges in Josephine, Jackson, Baker and Grant counties, as well as the economic factors that impacted this mining method. This volume also offers a unique look into the values of river bottom land in relation to both farming and mining, in how farm lands were mined, re-soiled and reclamated after the dredges worked them. Featured are hard to find maps of the gold dredge fields, as well as rare photographs from a bygone era. **8.5" X 11", 86 ppgs. Retail Price: $8.99**

Quick Silver Mining in Oregon - Originally published in 1963, this important publication on Oregon Mining has not been available for over fifty years. This publication includes details into the history and production of Elemental Mercury or Quicksilver in the State of Oregon. **8.5" X 11", 238 ppgs. Retail Price: $15.99**

Mines of the Greenhorn Mining District of Grant County Oregon - Originally published in 1948, this important publication on Oregon Mining has not been available for over sixty five years. In this publication are rare insights into the mines of the famous Greenhorn Mining District of Grant County, Oregon, especially the famous Morning Mine. Also included are details on the Tempest, Tiger, Bi-Metallic, Windsor, Psyche, Big Johnny, Snow Creek, Banzette and Paramount Mines, as well as prospects in the vicinities in the famous mining areas of Mormon Basin, Vinegar Basin and Desolation Creek. Included are hard to find mine maps and dozens of rare photographs from the bygone era of Grant County's rich mining history. **8.5" X 11", 72 ppgs. Retail Price: $9.99**

Geology of the Wallowa Mountains of Oregon: Part I (Volume 1) - Originally published in 1938, this important publication on Oregon Mining has not been available for nearly seventy five years. Included are details on the geology of this unique portion of North Eastern Oregon. This is the first part of a two book series on the area. Accompanying the text are rare photographs and historic maps.**8.5" X 11", 92 ppgs. Retail Price: $9.99**

Geology of the Wallowa Mountains of Oregon: Part II (Volume 2) - Originally published in 1938, this important publication on Oregon Mining has not been available for nearly seventy five years. Included are details on the geology of this unique portion of North Eastern Oregon. This is the first part of a two book series on the area. Accompanying the text are rare photographs and historic maps.**8.5" X 11", 94 ppgs. Retail Price: $9.99**

Field Identification of Minerals For Oregon Prospectors - Originally published in 1940, this important publication on Oregon Mining has not been available for nearly seventy five years. Included in this volume is an easy system for testing and identifying a wide range of minerals that might be found by prospectors, geologists and rockhounds in the State of Oregon, as well as in other locales. Topics include how to put together your own field testing kit and how to conduct rudimentary tests in the field. This volume is written in a clear and concise way to make it useful even for beginners. **8.5" X 11", 158 ppgs. Retail Price: $14.99**

The Bohemia Mining District of Oregon - Originally published in 1900, this important publication on Oregon Mining has not been available for over a century. Included in this volume are important insights into the famous Bohemia Mining District of Oregon, including the histories and locations of important gold mines in the area such as the Ophir Mine, Clarence, Acturas, Peek-a-boo, White Swan, Combination Mine, the Musick Mine, The California, White Ghost, The Mystery, Wall Street, Vesuvius, Story, Lizzie Bullock, Delta, Elsie Dora, Golden Slipper, Broadway, Champion Mine, Knott, Noonday, Helena, White Wings, Riverside and others. Also included are notes on the nearby Blue River Mining District. **8.5" X 11", 58 ppgs. Retail Price: $9.99**

The Gold Fields of Eastern Oregon - Unavailable since 1900, this publication was originally compiled by the Baker City Chamber of Commerce Offering important insights into the gold mining history of Eastern Oregon, "The Gold Fields of Eastern Oregon" sheds a rare light on many of the gold mines that were operating at the turn of the 19th Century in Baker County and Grant County in North Eastern Oregon. Some of the areas featured include the Cable Cove District, Baisely-Elhorn, Granite, Red Boy, Bonanza, Susanville, Sparta, Virtue, Vaughn, Sumpter, Burnt River, Rye Valley and other mining districts. Included is basic information on not only many gold mines that are well known to those interested in Eastern Oregon mining history, but also many mines and prospects which have been mostly lost to the passage of time. Accompanying are numerous rare photos **8.5" X 11", 78 ppgs. Retail Price: $10.99**

Gold Mining in Eastern Oregon - Originally published in 1938, this important publication on Oregon Mining has not been available for over a century. Included in this volume are important insights into the famous mining districts of Eastern Oregon during the late 1930's. Particular attention is given to those gold mines with milling and concentrating facilities in the Greenhorn, Red Boy, Alamo, Bonanza, Granite, Cable Cove, Cracker Creek, Virtue, Keating, Medical Springs, Sanger, Sparta, Chicken Creek, Mormon Basin, Connor Creek, Cornucopia and the Bull Run Mining Districts. Some of the mines featured include the Ben Harrison, North Pole-Columbia, Highland Maxwell, Baisley-Elkhorn, White Swan, Balm Creek, Twin Baby, Gem of Sparta, New Deal, Gleason, Gifford-Johnson, Cornucopia, Record, Bull Run, Orion and others. Of particular interest are the mill flow sheets and descriptions of milling operations of these mines. 8.5" X 11", 68 ppgs. Retail Price: $8.99

The Gold Belt of the Blue Mountains of Oregon - Originally published in 1901, this important publication on Oregon Mining has not been available for over a century. Included in this volume are rare insights into the gold deposits of the Blue Mountains of North East Oregon, including the history of their early discovery and early production. Extensive details are offered on this important mining area's mineralogy and economic geology, as well as insights into nearby gold placers, silver deposits and copper deposits. Featured are the Elkhorn and Rock Creek mining districts, the Pocahontas district, Auburn and Minersville districts, Sumpter and Cracker Creek, Cable Cove, the Camp Carson district, Granite, Alamo, Greenhorn, Robinsonville, the Upper Burnt River Valley and Bonanza districts, Susanville, Quartzburg, Canyon Creek, Virtue, the Copper Butte district, the North Powder River, Sparta, Eagle Creek, Cornucopia, Pine Creek, Lower Powder River, the Upper Snake River Canyon, Rye Valley, Lower Burnt River Valley, Mormon Basin, the Malheur and Clarks Creek districts, Sutton Creek and others. Of particular interest are important details on numerous gold mines and prospects in these mining districts, including their locations, histories, geology and other important information, as well as information on silver, copper and fire opal deposits. **8.5" X 11", 250 ppgs. Retail Price: $24.99**

Mining in the Cascades Range of Oregon - Originally published in 1938, this important publication on Oregon Mining has not been available for over seventy five years. Included in this volume are rare insights into the gold mines and other types of metal mines in the Cascades Mountain Range of Oregon. Some of the important mining areas covered include the famous Bohemia Mining District, the North Santiam Mining District, Quartzville Mining District, Blue River Mining District, Fall Creek Mining District, Oakridge District, Zinc District, Buzzard-Al Sarena District, Grand Cove, Climax District and Barron Mining District. Of particular interest are important details on over 100 mines and prospects in these mining districts, including their locations, histories, geology and other important information. **8.5" X 11", 170 ppgs. Retail Price: $14.99**

Beach Gold Placers of the Oregon Coast - Originally published in 1934, this important publication on Oregon Mining has not been available for over 80 years. Included in this volume are rare insights into the beach gold deposits of the State of Oregon, including their locations, occurance, composition and geology. Of particular interest is information on placer platinum in Oregon's rich beach deposits. Also included are the locations and other information on some famous Oregon beach mines, including the Pioneer, Eagle, Chickamin, Iowa and beach placer mines north of the mouth of the Rogue River. **8.5" X 11", 60 ppgs. Retail Price: $8.99**

Mineralogical Composition of the Sands of the Oregon Coast: From Coos Bay to the Columbia - Published in 1945, he text features hard to find information on the composition of the gold bearing black sands of the South West Oregon Coast, offering a unique insight to prospectors in search of Oregon's legendary beach gold. 104 ppgs, $9.99

Manganese Mining in Oregon - First released in 1942 and now out of print, this special reprint edition of "Manganese in Oregon" was originally published by the Oregon Department of Geology and Mineral Industries. The text features hard to find information on the mining of Manganese in Oregon, including details and maps of Oregon manganese mines and prospects. 108 ppgs, 9.99

Medford Oregon As A Mining Center - Written in 1912, this hard to find publication includes valuable insights into the mining history of South West Oregon. This small book contains interesting information on the gold, copper and mining industry in Southern Oregon as it existed just prior to World War One, shedding light on some of the important mines in the area. Included are rare photographs and vintage advertising of the day. 80 ppgs, 9.99

Mineral Resources of Curry County Oregon - First released in 1977 and now out of print, this special reprint edition of "Geology, Mineral Resources and Rock Materials of Curry County, Oregon" was originally published in cooperation of Curry County, Oregon and the Oregon Department of Geology and Mineral Industries. The text features hard to find information on not only the mining of gold and other metals in Curry County, but also aggregate mining in the area. 102 ppgs, 11.99

Origin of the Gold Bearing Black Sands of the Coast of South West Oregon - First released in 1943 and now out of print, this special reprint edition of "The Origin of the Black Sands of the South West Oregon Coast" was originally published by the Oregon Department of Geology and Mineral Industries. The text features hard to find information on the origin of the gold bearing black sands of the South West Oregon Coast, offering a unique insight to prospectors in search of Oregon's legendary beach gold. 52 ppgs, 8.99

South West Oregon Mining - Leading mining historian Kerby Jackson introduces us to six classic small mining publications on the Gold Mining Industry in Southern Oregon. This small book consists of a compilation of USGS J.S. Diller's "Mines of the Riddles Quadrangle", "The Rogue River Valley Coal Fields" and "Mineral Resources of the Grants Pass Quadrangle", the Grants Pass Commercial Club's rare publication "Mining in Josephine County, Oregon" and the USGS publication "The Distribution of Placer Gold in the Sixes River, South West Oregon". Also included is F.W. Libbey's legendary article on the Southern Oregon Mining Industry, "Lest We Forget", which appeared in the publication of the Oregon State Department of Geology and Mineral Industries in the early 1960's. This compilation offers a unique perspective on mining in South West Oregon and includes considerable information on mines in Josephine, Jackson and Coos Counties. 142 ppgs, 14.99

Geology and Mineral Resources of the Gasquet Quadrangle of California-Oregon - First published in 1953, it has been unavailable for over a century and sheds important light on the geological features and mineral resources of this portion of Northern California and Southern Oregon. 80 ppgs, 9.99

The Little North Santiam Mining District of Oregon - Unavailable since 1985, this publication offers rare insights into one of the most famous mining areas in Western Oregon. Of special interest is this publication's focus on the history of the most important gold mines in the Little North Santiam Mining District. Illustrated with hard to find historical photos. 102 ppgs, 14.99

The Economic Geological Resources of Oregon - Unavailable since 1912, this publication offers rare insights into the early history of mining in Oregon. Included is hard to find information on gold, silver, copper and other mines that operated in Oregon at the turn of the century. 126 ppgs, 14.99

Sights in the Gold Region of Oregon and California - Unavailable since 1853, this publication provides a fascinating insight into the California and Oregon Gold Rushes through the eyes of one of the men who went West and "saw the elephant" to take part in it. Theodore Taylor Johnson's memoir of his journey to the gold fields of California and Oregon offers a unique look into this important time during the settling of the Far West. 382 ppgs, 24.99

South Dakota Mining Books

Mining and Metallurgy of the Black Hills of South Dakota - Mining historian Kerby Jackson introduces us to a classic mining work in this important re-issue of "Papers Read Before The Black Hills Mining Men's Association At Their Regularly Monthly Meeting On The Mining and Metallurgy of the Black Hills Ores". Unavailable since 1904, this publication offers rare insights into the famous bLack Hills mining region of South Dakota. Topics include Mining and Milling Methods of the Black Hills, South Dakota Gold Production, Some Features of the Mining Operations in the Homestake Mine at Lead, South Dakota, The Metallurgy of the Ore in the Homestake Mine, Cyanidation of Black Hills Ores, Wet Crushing of Ores in Solution, Cyaniding Practices at the Maitland Mine, Pyrite Ores and Their Smelting, Matte Smelting, Mining in the Bald Mountain and Ruby Districts of the Black Hills of South Dakota and more. Lavishly illustrated with rare historical photographs. 8.5" X 11", 162 ppgs, Retail Price: $14.99

Utah Mining Books

Fluorite in Utah - Unavailable since 1954, this publication was originally compiled by the USGS, State of Utah and U.S. Atomic Energy Commission and details the mining of fluorspar, also known as fluorite in the State of Utah. Included are details on the geology and history of fluorspar (fluorite) mining in Utah, including details on where this unique gem mineral may be found in the State of Utah. 8.5" X 11", 60 ppgs. Retail Price: $8.99

The Gold Hill Mining District of Utah - First published in 1935, it has been unavailable since those days and sheds important light on the mines, history and geology of Utah's Gold Hill Mining District. Included are rare insights into this important mining area, including the locations, histories and details of numerous mines. This volume is well illustrated with geological diagrams, as well as hard to find maps of some of the most important mines in this district. 202 ppgs., 19.99

The Mines, Miners and Minerals of Utah - First published in 1896, it has been unavailable since those days and sheds important light on the early mines and miners of Pioneer Utah, as well as the minerals which they won from the earth by laborious hard physical labor and sheer determination. Included are rare insights into the early mining history of Utah, as well details on hundreds of gold, silver and copper mines. 376 ppgs., 24.99

Washington Mining Books

The Republic Mining District of Washington - Unavailable since 1910, this important publication was originally published by the Washington Geologic Survey and has been unavailable for a century. Topics include the geology, rock formations and the formation of ore deposits in this important mining area of Washington State. Also included are hard to find details on the geology, history and locations of dozens of mines in the area. Some of the mines featured include the New Republic Mine, Ben Hur, Morning Glory, the South Republic Mine, Quilp, Surprise, Black Tail, Lone Pine, San Poil, Mountain Lion, Tom Thumb, Elcaliph and many others. 8.5" X 11", 94 ppgs, Retail Price: $10.99

The Myers Creek and Nighthawk Mining Districts of Washington - Unavailable since 1911, this important publication was originally published by the Washington Geologic Survey and has been unavailable for a century. Topics include the geology, rock formations and the formation of ore deposits in these important mining areas of Washington State. Also included are hard to find details on the geology, history and locations of dozens of mines in the area. Some of the mines featured include the Grant Mine, Monterey, Nip and Tuck, Myers Creek, Number Nine, Neutral, Rainbow, Aztec, Crystal Butte, Apex, Butcher Boy, Molson, Mad River, Olentangy, Delate, Kelsey, Golden Chariot, Okanogan, Ohio, Forty-Ninth Parallel, Nighthawk, Favorite, Little Chopaka, Summit, Number One, California, Peerless, Caaba, Prize Group, Ruby, Mountain Sheep, Golden Zone, Rich Bar, Similkameen, Kimberly, Triune, Hiawatha, Trinity, Hornsilver, Maquae, Bellevue, Bullfrog, Palmer Lake, Ivanhoe, Copper World and many others.
8.5" X 11", 136 ppgs, Retail Price: $12.99

The Blewett Mining District of Washington - Unavailable since 1911, this important publication was originally published by the Washington Geologic Survey and has been unavailable for a century. Topics include the geology, rock formations and the formation of ore deposits in this important mining area of Washington State. Also included are hard to find details on the geology, history and locations of dozens of mines in the area. Some of the mines featured include the Washington Meteor, Alta Vista, Pole Pick, Blinn, North Star, Golden Eagle, Tip Top, Wilder, Golden Guinea, Lucky Queen, Blue Bell, Prospect, Homestake, Lone Rock, Johnson, and others. **8.5" X 11", 134 ppgs, Retail Price: $12.99**

Silver Mining In Washington - Unavailable since 1955, this important publication was originally published by the Washington Geologic Survey. Featured are the hard to find locations and details pertaining to Washington's silver mines. **8.5" X 11", 180 ppgs, Retail Price: $15.99**

The Mines of Snohomish County Washington - Unavailable since 1942, this important publication was originally published by the Washington Geologic Survey and has been unavailable for seventy years. Featured are details on a large number of gold, silver, copper, lead and other metallic mineral mines. Included are the locations of each historic mine, along with information on the commodity produced. **8.5" X 11", 98 ppgs, Retail Price: $10.99**

The Mines of Chelan County Washington - Unavailable since 1943, this important publication was originally published by the Washington Geologic Survey and has been unavailable for seventy years. Featured are details on a large number of gold, silver, copper, lead and other metallic mineral mines. Included are the locations of each historic mine, along with information on the commodity. **8.5" X 11", 88 ppgs, Retail Price: $9.99**

Metal Mines of Washington - Unavailable since 1921, this important publication was originally published by the Washington Geologic Survey and has been unavailable for nearly ninety years. Widely considered a masterpiece on the Washington Mining Industry, "Metal Mines of Washington" sheds light on the important details of Washington's early mining years. Featured are details on hundreds of gold, silver, copper, lead and other metallic mineral mines. Included are hard to find details on the mineral resources of this state, as well as the locations of historic mines. Lavishly illustrated with maps and historic photos and complete with a glossary to explain any technical terms found in the text, this is one of the most important works on mining in the State of Washington. No prospector or miner should be without it if they are interested in mining in Washington. **8.5" X 11", 396 ppgs, Retail Price: $24.99**

Gem Stones In Washington - Unavailable since 1949, this important publication was originally published by the Washington Geologic Survey and has been unavailable since first published. Included are details on where to find naturally occurring gem stones in the State of Washington, including quartz crystal, amethyst, smoky quartz, milky quartz, agates, bloodstone, carnelian, chert, flint, jasper, onyx, petrified wood, opal, fire opal, hyalite and others. **8.5" X 11", 54 ppgs, Retail Price: $8.99**

The Covada Mining District of Washington - Unavailable since 1913, this important publication was originally published by the Washington Geologic Survey and has been unavailable for a century. Topics include the geology, rock formations and the formation of ore deposits in this important mining area of Washington State. Also included are hard to find details on the geology, history and locations of dozens of mines in the area. Some of the mines featured include the Admiral, Advance, Algonkian, Big Bug, Big Chief, Big Joker, Black Hawk, Black Tail, Black Thorn, Captain, Cherokee Strip, Colorado, Dan Patch, Dead Shot, Etta, Good Ore, Greasy Run, Great Scott, Idora, IXL, Jay Bird, Kentucky Bell, King Solomon, Laurel, Laura S, Little Jay, Meteor, Neglected, Northern Light, Old Nell, Plymouth Rock, Polaris, Quandary, Reserve, Shoo Fly, Silver Plume, Three Pines, Vernie, White Rose and dozens of others. **8.5" X 11", 114 ppgs, Retail Price: $10.99**

The Index Mining District of Washington - Unavailable since 1912, this important publication was originally published by the Washington Geologic Survey and has been unavailable for a century. Topics include the geology, rock formations and the formation of ore deposits in this important mining area of Washington State. Also included are hard to find details on the geology, history and locations of dozens of mines in the area. Some of the mines featured include the Sunset, Non-Pareil, Ethel Consolidated, Kittaning, Merchant, Homestead, Co-operative, Lost Creek, Uncle Sam, Calumet, Florence-Rae, Bitter Creek, Index Peacock, Gunn Peak, Helena, North Star, Buckeye. Copper Bell, Red Cross and others. **8.5" X 11", 114 ppgs, Retail Price: $11.99**

Mining & Mineral Resources of Stevens County Washington - Unavailable since 1920, this important publication was originally published by the Washington Geologic Survey and has been unavailable for a century. Topics include the geology, rock formations and the formation of ore deposits in these important mining areas of Washington State. Also included are hard to find details on the geology, history and locations of hundreds of mines in the area. **8.5" X 11", 372 ppgs, Retail Price: $24.99**

The Mines and Geology of the Loomis Quadrangle Okanogan County, Washington - Unavailable since 1972, this important publication was originally published by the Washington Geologic Survey and has been unavailable for a century. Topics include the geology, rock formations and the formation of ore deposits in this important mining area of Washington State. Also included are hard to find details on the geology, history and locations of dozens of gold, copper, silver and other mines in the area. **8.5" X 11", 150 ppgs, Retail Price: $12.99**

The Conconully Mining District of Okanogan County Washington - Unavailable since 1973, this important publication was originally published by the Washington Geologic Survey and has been unavailable for a century. Topics include the geology, rock formations and the formation of ore deposits in this important mining area of Washington State, which also includes Salmon Creek, Blue Lake and Galena. Also included are hard to find details on the geology, mining history and locations of dozens of mines in the area. Some of the mines include Arlington, Fourth of July, Sonny Boy, First Thought, Last Chance, War Eagle-Peacock, Wheeler, Mohawk, Lone Star, Woo Loo Moo Loo, Keystone, Hughes, Plant-Callahan, Johnny Boy, Leuena, Gubser, John Arthur, Tough Nut, Homestake, Key and many others **8.5" X 11", 68 ppgs, Retail Price: $8.99**

Gold Hunting in the Cascade Mountains of Washington - First published in 1861, this rare publication offers rare insights into an early search for placer gold near Mount Saint Helens in what was then Washington Territory. This rare booklet was written by an anonymous author under the name Loo-Wit Lat-Kla, which is a Native American word for "fire mountain", referring to Mount St. Helens. Gold Hunting in the Cascade Mountains is a fascinating read on the early history of mining in Washington, as well as on the mountaineering of Mount St. Helens. Only one copy of the original text survives. In the 1950's a limited edition of 300 copies was produced by Yale University, few of which still survive today. **8.5" X 11", 56 ppgs, Retail Price: $8.99**

Wyoming Mining Books

Mining in the Laramie Basin of Wyoming - Unavailable since 1909, this publication was originally compiled by the United States Department of Interior. Also included are insights into the mineralization and other characteristics of this important mining region, especially in regards to coal, limestone, gypsum, bentonite clay, cement, sand, clay and copper. **8.5" X 11", 104 ppgs, Retail Price: $11.99**

More Mining Books

Prospecting and Developing A Small Mine - Topics covered include the classification of varying ores, how to take a proper ore sample, the proper reduction of ore samples, alluvial sampling, how to understand geology as it is applied to prospecting and mining, prospecting procedures, methods of ore treatment, the application of drilling and blasting in a small mine and other topics that the small scale miner will find of benefit. **8.5" X 11", 112 ppgs, Retail Price: $11.99**

Timbering For Small Underground Mines - Topics covered include the selection of caps and posts, the treatment of mine timbers, how to install mine timbers, repairing damaged timbers, use of drift supports, headboards, squeeze sets, ore chute construction, mine cribbing, square set timbering methods, the use of steel and concrete sets and other topics that the small underground miner will find of benefit. This volume also includes twenty eight illustrations depicting the proper construction of mine timbering and support systems that greatly enhance the practical usability of the information contained in this small book. **8.5" X 11", 88 ppgs. Retail Price: $10.99**

Timbering and Mining - A classic mining publication on Hard Rock Mining by W.H. Storms. Unavailable since 1909, this rare publication provides an in depth look at American methods of underground mine timbering and mining methods. Topics include the selection and preservation of mine timbers, drifting and drift sets, driving in running ground, structural steel in mine workings, timbering drifts in gravel mines, timbering methods for driving shafts, positioning drill holes in shafts, timbering stations at shafts, drainage, mining large ore bodies by means of open cuts or by the "Glory Hole" system, stoping out ore in flat or low lying veins, use of the "Caving System", stoping in swelling ground, how to stope out large ore bodies, Square Set timbering on the Comstock and its modifications by California miners, the construction of ore chutes, stoping ore bodies by use of the "Block System", how to work dangerous ground, information on the "Delprat System" of stoping without mine timbers, construction and use of headframes and much more. This volume provides a reference into not only practical methods of mining and timbering that may be employed in narrow vein mining by small miners today, but also rare insights into how mines were being worked at the turn of the 19th Century. **8.5" X 11", 288 ppgs. Retail Price: $24.99**

A Study of Ore Deposits For The Practical Miner - Mining historian Kerby Jackson introduces us to a classic mining publication on ore deposits by J.P. Wallace. First published in 1908, it has been unavailable for over a century. Included are important insights into the properties of minerals and their identification, on the occurrence and origin of gold, on gold alloys, insights into gold bearing sulfides such as pyrites and arsenopyrites, on gold bearing vanadium, gold and silver tellurides, lead and mercury tellurides, on silver ores, platinum and iridium, mercury ores, copper ores, lead ores, zinc ores, iron ores, chromium ores, manganese ores, nickel ores, tin ores, tungsten ores and others. Also included are facts regarding rock forming minerals, their composition and occurrences, on igneous, sedimentary, metamorphic and intrusive rocks, as well as how they are geologically disturbed by dikes, flows and faults, as well as the effects of these geologic actions and why they are important to the miner. Written specifically with the common miner and prospector in mind, the book will help to unlock the earth's hidden wealth for you and is written in a simple and concise language that anyone can understand. **8.5" X 11", 366 ppgs. Retail Price: $24.99**

Mine Drainage - Unavailable since 1896, this rare publication provides an in depth look at American methods of underground mine drainage and mining pump systems. This volume provides a reference into not only practical methods of mining drainage that may be employed in narrow vein mining by small miners today, but also rare insights into how mines were being worked at the turn of the 19th Century. **8.5" X 11", 218 ppgs. Retail Price: $24.99**

Fire Assaying Gold, Silver and Lead Ores - Unavailable since 1907, this important publication was originally published by the Mining and Scientific Press and was designed to introduce miners and prospectors of gold, silver and lead to the art of fire assaying. Topics include the fire assaying of ores and products containing gold, silver and lead; the sampling and preparation of ore for an assay; care of the assay office, assay furnaces; crucibles and scorifiers; assay balances; metallic ores; scorification assays; cupelling; parting' crucible assays, the roasting of ores and more. This classic provides a time honored method of assaying put forward in a clear, concise and easy to understand language that will make it a benefit to even beginners. **8.5" X 11", 96 ppgs. Retail Price: $11.99**

Methods of Mine Timbering - Originally published in 1896, this important publication on mining engineering has not been available for nearly a century. Included are rare insights into historical methods of timbering structural support that were used in underground metal mines during the California that still have a practical application for the small scale hardrock miner of today. **8.5" X 11", 94 ppgs. Retail Price: $10.99**

The Enrichment of Copper Sulfide Ores - First published in 1913, it has been unavailable for over a century. Topics include the definition and types of ore enrichment, the oxidation of copper ores, the precipitation of metallic sulfides. Also included are the results of dozens of lab experiments pertaining to the enrichment of sulfide ores that will be of interest to the practical hard rock mine operator in his efforts to release the metallic bounty from his mine's ore. **8.5" X 11", 92 ppgs. Retail Price: $9.99**

A Study of Magmatic Sulfide Ores - Unavailable since 1914, this rare publication provides an in depth look at magmatic sulfide ores. Some of the topics included are the definition and classification of magmatic ores, descriptions of some magmatic sulfide ore deposits known at the time of publication including copper and nickel bearing pyrrohitic ore bodies, chalcopyrite-bornite deposits, pyritic deposits, magnetite-ileminite deposits, chromite deposits and magmatic iron ore deposits. Also included are details on how to recognize these types of ore deposits while prospecting for valuable hardrock minerals. **8.5" X 11", 138 ppgs. Retail Price: $11.99**

The Cyanide Process of Gold Recovery - Unavailable since 1894 and released under the name "The Cyanide Process: Its Practical Application and Economical Results", this rare publication provides an in depth look at the early use of cyanide leaching for gold recovery from hardrock mine ores. This volume provides a reference into the early development and use of cyanide leaching to recover gold. **8.5" X 11", 162 ppgs. Retail Price: $14.99**

California Gold Milling Practices - Unavailable since 1895 and released under the name "California Gold Practices", this rare publication provides an in depth look at early methods of milling used to reduce gold ores in California during the late 19th century. This volume provides a reference into the early development and use of milling equipment during the earliest years of the California Gold Rush up to the age of the Industrial Revolution. Much of the information still applies today and will be of use to small scale miners engaging in hardrock mining. **8.5" X 11", 104 ppgs. Retail Price: $10.99**

Leaching Gold and Silver Ores With The Plattner and Kiss Processes - Mining historian Kerby Jackson introduces us to a classic mining publication on the evaluation and examination of mines and prospects by C.H. Aaron. First published in 1881, it has been unavailable for over a century and sheds important light on the leaching of gold and silver ores with the Plattner and Kiss processes. **8.5" X 11", 204 ppgs. Retail Price: $15.99**

The Metallurgy of Lead and the Desilverization of Base Bullion - First published in 1896, it has been unavailable for over a century and sheds important light on the the recovery of silver from lead based ores. Some of the topics include the properties of lead and some of its compounds, lead ores such as galenite, anglesite, cerussite and others, the distribution of lead ores throughout the United States and the sampling and assaying of lead ores. Also covered is the metallurgical treatment of lead ores, as well as the desilverization of lead by the Pattinson Process and the Parkes Process. Hofman's text has long been considered one of the most important early works on the recovery of silver from lead based ores. 8.5" X 11", 452 ppgs. Retail Price: $29.99

Ore Sampling For Small Scale Miners - First published in 1916, it has been unavailable for over a century and sheds important light on historic methods of ore sampling in hardrock mines. Topics include how to take correct ore samples and the conditions that affect sampling, such as their subdivision and uniformity. Particular detail is given to methods of hand sampling ore bodies by grab sample, pipe sample and coning, as well as sampling by mechanical methods. Also given are insights into the screening, drying and grinding processes to achieve the most consistent sample results and much more. 8.5" X 11", 124 ppgs. Retail Price: $12.99

The Extraction of Silver, Copper and Tin from Ores - First published in 1896, it has been unavailable for over a century and sheds important light on how historic miners recovered silver, copper and tin from their mining operations. The book is split into three sections, including a discussion on the Lixiviation of Silver Ores, the mining and treatment of copper ores as practiced at Tharsis, Spain and the smelting of tin as it was practiced by metallurgists at Pulo Brani, Singapore. Also included is an overview and analysis of these historic metal recovery methods that will be of benefit to those interested in the extraction of silver, copper and tin from small mines. 8.5" X 11", 118 ppgs. Retail Price: $14.99

The Roasting of Gold and Silver Ores - First published in 1880, it has been unavailable for over a century and sheds important light on how historic miners recovered gold and silver rom their mining operations. Topics include details on the most important silver and free milling gold ores, methods of desulphurization of ores, methods of deoxidation, the chlorination of ores, methods and details on roasting gold and silver ores, notes on furnaces and more. Also included are details on numerous methods of gold and silver recovery, including the Ottokar Hofman's Process, the Patera Process, Kiss Process, Augustin Process, Ziervogel Process and others. 8.5" X 11", 178 ppgs. Retail Price: $19.99

The Examination of Mines and Prospects - First published in 1912, it has been unavailable for over a century and sheds important light on how to examine and evaluate hardrock mines, prospects and lode mining claims. Sections include Mining Examinations, Structural Geology, Structural Features of Ore Deposits, Primary Ores and their Distribution, Types of Primary Ore Deposits, Primary Ore Shoots, The Primary Alteration of Wall Rocks, Alterations by Surface Agencies, Residual Ores and their Distribution, Secondary Ores and Ore Shoots and Vein Outcrops. This hard to find information is a must for those who are interested in owning a mine or who already own a lode mining claim and wish to succeed at quartz mining. 8.5" X 11", 250 ppgs. Retail Price: $19.99

Garnets: Their Mining, Milling and Utilization - First published in 1925, it has been unavailable since those days and sheds important light on the mining, milling and utilization of garnets. Included are details on the characteristics of garnets, where they are found and how they were mined. 78 ppgs, 10.99

Gemstones and Precious Stones of North America - Leading mining historian Kerby Jackson introduces us to a classic mining publication on the gems and precious stones of the United States, Canada and mexico. First published in 1890, it has been unavailable since those days and sheds important light on the gems and precious stones that may be found in North America. Included are chapters on diamonds, corundum, sapphire, ruby, topaz, emerald, disapore, spinel, turquoise, tourmaline, garnets, beyrl, peridot, zircon, quartz crystals, feldspars, pearls and many others. Included are details on where these gems and precious stones may be found throughout North America, as well as their characteristics. 360 ppgs, 24.99

Mining Camps and Mining Districts - First released in 1885 by Charles Howard Shinn under the title "Mining Camps: A Study in American Frontier Government", this publication offers a unique look at how early gold miners established their own forms of representative government during the California Gold Rush. Drawing on the the early mining codes of mideviel German miners in the Harz Mountains, on the mining customs of the Cornish tin miners and early Spanish mining laws introduced into California, the miners established the first governments in the American West. 340 ppgs, 24.99

BLM Field Handbook for Mineral Examiners - Leading mining historian Kerby Jackson introduces us to a classic mining publication on mine evaluation. First published in 1962, this work sheds important light on the techniques of BLM Mineral Examiners to perform validity on mining claims. 132 ppgs, 10.99

<u>Six Months In The Gold Mines During The California Gold Rush</u> - Unavailable since 1850, this important work is a first hand account of one "49'ers" personal experience during the great California Gold Rush, shedding important light on one of the most exciting periods in the history of not only California, but also the world. Compiled from journals written between 1847 and 1849 by E. Gould Buffum, a native of New York, "Six Months In The Gold Mines During The California Gold Rush" offers a rare look into the day to day lives of the people who came to California to work in her gold mines when the state was still a great frontier. 8.5" X 11", 290 ppgs. Retail Price: $19.99

<u>The Discovery of Gold in Australia</u> - First published in 1852, it has been unavailable since those days and sheds important light on Australia's gold mining history. Included are rare communications between British agents and the British Crown when gold was first discovered in Australia in 1851. This rare text contains hard to find details on Australia's first mining camps and Britain's early attempts to provide for the orderly regulation of gold mines in that part of the world. Also of interest are hard to find extracts of articles that appeared in the early colonial newspapers that did their best to report on Australia's gold rush as it took place.
102 ppgs, 10.99

<u>Notes on Ore Sampling in Mines</u> - Unavailable since 1903, this publication offers rare insights into how ore was sampled in metallic mineral mines at the turn of the 19th Century. Included in this small booklet are facts about how to take, separate and handle an ore sample, as well as details on some of the equipment that was used for sampling in the old days.
68 ppgs, 7.99

<u>Elementary Methods of Placer Gold Mining</u> - Unavailable since 1944, this publication offers rare insights into the art of finding and recovering placer gold. Included in this small booklet are facts about the geology of alluvial gold deposits, the various types of placer gold deposits and the metals associated with placer gold. Also included are basic instructions on panning for gold, the use of sluice boxes, rocker boxes, as well as the recovery of fine gold by amalgamation plates and other methods. Basic plans to build your own mining equipment is also included. Written mainly for miners in Idaho, this short booklet also includes an overview of where to find gold in Idaho.
58 ppgs, 7.99

<u>Mining Districts of the Western United States</u> - Unavailable since 1912, this publication provides the locations and other basic information on the mining districts of the Western United States. This important reference book provides valuable insights into the general locations of where gold, silver, copper and other mines have operated in the Western States. This fascinating book offers a rare glimpse into these marvels of early mining technology that once helped early miners win millions of ounces of gold and silver from the hills of the Far West. 336 ppgs, 24.99

<u>Some Facts About Ore Deposits</u> - Written in 1935, this hard to find publication includes valuable facts on the nature of metallic ore deposits. Highlighted here are the details on how ores are deposited, on the fallacy that ore deposits always increase in value with depth, primary ore zones, myths regarding the leaching of ores, facts about secondary ore enrichment, which rocks are associated with which types of metals and much more. This small booklet will be found to be of immense value to the miner who is looking to learn about hard rock mining. 126 ppgs, 11.99

<u>Prospecting Field Tests For The Common Metals</u> - Written in 1942, this hard to find publication includes valuable facts on how to identify common metals in the field. Included are field tests for gold, silver, copper, arsenic, antimony, iron, chromium, manganese, lead, cobalt, nickel, tin, tungsten, zinc, vanadium and many other minerals utilizing reagents, blowpipes and other methods. This small booklet will be found to be of immense value to the miner who is looking to learn about hard rock mining. 82 ppgs, 8.99

<u>Sampling for Gold</u> - Leading mining historian Kerby Jackson brings together five historic publications from the Arizona Bureau of Mines on the subject of sampling and testing for gold, be it placer or lode gold. Included in this publication are "Mill and Smelter Methods of Sampling" (1915), "Sampling and the Estimation of Gold in a Placer Deposit" (1917), "Sampling of Ore Dumps and Tailings" (1917), "Sampling Mineralized Veins" (1918) and "Select Blowpipe and Acid Tests for Minerals" (1918). As sampling is the most important activity that a miner or prospector seeking gold needs to engage in, these tried and proven methods of sampling will be found to greatly assist those seeking their own golden fortune. 86 ppgs, 10.99

<u>Treating Gold Ores</u> - Written in 1932, this hard to find publication includes valuable facts about the handling of ores from gold mines. Included in this short publication is an overview of smelting, milling, amalgamation, gravity stamp milling, the use of retorts, the refining of bullion from retorts, use of ball mills, huntington mills and arrastras, as well as details on cyanidation, gravity concentration and flotation. This publication is a must for anyone looking to develop a small gold mine. 90 ppgs, 9.99

<u>Selling Mines and Prospects</u> - Leading mining historian Kerby Jackson introduces us to a classic mining publication on the selling of mines and prospects. Written in 1918, this hard to find publication includes valuable facts about how mines and prospects were sold in decades past that will still be found to be of use today. 46 ppgs, 7.99

<u>Mining Stamp Mills</u> - Unavailable since 1912, this publication offers rare insights into the development and use of stamp mills that were once employed in gold and other mines in the century past. Included are details on the history of stamp mills, including their evolution from the Cornish Mill, Appalachian Mill and the California Mill, as well as the construction and operation of these mills in mining operations. This fascinating book offers a rare glimpse into these marvels of early mining technology that once helped early miners win millions of ounces of gold and silver from the hills of the Far West. 164 ppgs, 14.99

<u>Notes on Ore Sampling in Mines</u> - Unavailable since 1903, this publication offers rare insights into how ore was sampled in metallic mineral mines at the turn of the 19th Century. Included in this small booklet are facts about how to take, separate and handle an ore sample, as well as details on some of the equipment that was used for sampling in the old days. 68 ppgs, 7.99

www.ingramcontent.com/pod-product-compliance
Lightning Source LLC
Chambersburg PA
CBHW081152180526
45170CB00006B/2036